Religion and Nuclear Weapons:
A Study of Islamic Republic of Iran and Pakistan

Religion and Nuclear Weapons:
A Study of Islamic Republic of Iran and Pakistan

Shameer Modongal
Seyed Hossein Mousavian

Vij Books India Pvt Ltd

New Delhi (India)

Published by

Vij Books India Pvt Ltd
(Publishers, Distributors & Importers)
2/19, Ansari Road
Delhi – 110 002
Phones: 91-11-43596460
Mob: 9811094883
web: www.vijbooks.in
e-mail: contact@vijpublishing.com

ISBN: 978-93-93499-61-5   (Paperback)

Contents

List of Figures

List of Abbreviations

AEC	:	Atomic Energy Committee
AEOI	:	Atomic Energy Organization of Iran
BJP	:	Bharatiya Janata Party
CENTO	:	Central Treaty Organization
CIA	:	Central Intelligence Agency
CTBT	:	Comprehensive Test Ban Treaty
GCC	:	Gulf Cooperation Council
IAEA	:	International Atomic Energy Agency
INC	:	Indian National Congress
IR	:	International Relations
IRGC	:	Iranian Revolutionary Guards Corps
JCPOA	:	Joint Comprehensive Plan of Action
MIT	:	Massachusetts Institute of Technology
NIC	:	National Identity Conception
NIE	:	National Intelligence Estimate
NPT	:	Non-Proliferation Treaty
NWS	:	Nuclear Weapon States
OIC	:	Organization of Islamic Conference
PAEC	:	Pakistan Atomic Energy Commission
PNE	:	Peaceful Nuclear Explosion

PNE	:	Peaceful Nuclear Explosion
SEATO	:	Southeast Asia Treaty Organization
SNSC	:	Supreme National Security Council
WMD	:	Weapons of Mass Destruction

Introduction

This study explores the role of religion in shaping the nuclear policies of a state. It analyses whether religion becomes a motivating or de-motivating factor in the nuclear weapons programs. The fatwa of Ayatollah Khamenei, the Iranian Supreme Leader, against nuclear weapons is an important factor in the recent debate over the possibility of nuclear weaponization in Iran. At the same time, Pakistan's nuclear tests of 1998 were labelled as tests of an 'Islamic bomb'. In these two examples, religion is seen as a de-motivating factor for Iran, restricting the state from going into a nuclear weaponization program. In the case of Pakistan, the same religion is considered a motivating factor. However, mainstream theories about states' nuclear decision-making ignore the influence of religion. Therefore, a comprehensive and theoretical study is required to understand religion's role in states' nuclear policies.

Ample studies have analyzed the motivations behind nuclear weaponization and non-weaponization programs. Most of them, especially in traditional realist literature, connect nuclear policy with external threats to a state. Thus, the country is expected to seek nuclear weapons when it faces a threat to its security and to remain a non-nuclear-weapon state if it is not subject to a serious military threat. According to this view, as John M. Deutsch (1992) expressed, the "fundamental motivation to seek nuclear weapon is the perception that national security will be improved". Theories such as liberalism, neo-classical realism and constructivism have challenged this perspective and have argued that domestic politics is an important factor in policy-making of a state. The mere focus on the existence or absence of an external threat has been unable to explain the nuclear policies of many states, including Japan and France. Even though both were offered extended nuclear deterrence by the United States (US), France built and tested its nuclear weapons while Japan did not. As Scott Sagan explains, if the analysis is restricted to an external threat, the nuclear behaviour of India would be a puzzle since it did not test a nuclear weapon after its defeat by China but tested it after its victory over Pakistan. The absence

1

of proliferation by a large number of states, as would be predicted by the realist logic "proliferation begets proliferation", indicates the need to inquire into other factors affecting the nuclear policies of states.

Scott. D. Sagan (1996), in his famous article, "Why Do States Build Nuclear Weapons?: Three Models in Search of a Bomb," questions the realist view and presents three alternative theoretical frameworks to illustrate the need for a holistic approach. They include security, domestic politics, and the norms models. Thus Sagan suggests a multi-causality approach for explaining proliferation. Bhumitra Chakma (2002) puts forward four competing arguments for understanding why states go for nuclear weapons: (1) security concerns; (2) prestige; (3) technological imperatives; and (4) domestic politics. Bradley A. Thayer's (1995) "The Causes of Nuclear Proliferation and the Non-proliferation Regime," Stephen M. Meyer's (1984) *The Dynamics of Nuclear Proliferation*, and William Epstein's (1977) "Why States Go - And Don't Go – Nuclear" also deal with motivations behind the states' nuclear policies.

Though the literature on the incentives behind nuclear weaponization is plenty, they have largely ignored the impact of religion on nuclear policy. Even works which are based on a social-constructivist perspective, giving importance to social norms, do not assess the role of religion in the development of such norms in a given society. Sagan, for example, recognized the importance of norms, but he does not delve into the details of the role of religious norms or the role of religion in creating these norms.

After the Westphalia treaties of 1648, religion was separated from state, and the decisions of states were expected to be taken on the basis of national interests. Nonetheless, religion continues to be an important institution that influences day to day life of human beings, in both secular and theocratic states. The degree of its influence in policymaking varies from state to state. However, even in countries where religion has no direct impact on policies of the state, it influences the creation of norms and public opinion and thereby indirectly influences state policies. Thus, states sometimes make policies based on religious principles or justify their decisions taken for other interests by using religious texts and norms. The present study investigates religion's influence on a state's nuclear policies.

The book edited by Sohail H. Hashmi and Steven P. Lee (2004), *Ethics and Weapons of Mass Destruction: Religious and Secular Perspectives*, is one of the best books examining the views of different religions on WMD. In addition to the original debates of realist, liberalist and Christian views, the book examines Buddhist, Confucian, Hindu, Islamic and Judaic perspectives

2

on the ethics of WMD. In his chapter on "Islamic Ethics and Weapon of Mass Destruction: An Argument for Non-Proliferation", Hashmi divides the Islamic views on the ethics of WMD into three broad categories: WMD jihadists, WMD terrorists, and WMD pacifists. WMD terrorists and pacifists are at two opposing poles. While the former argues for the necessity of WMD, including nuclear weapons, the latter prohibits it even for deterrent purposes since it may escalate into actual war or is a wastage of resources without usage. WMD jihadists argue the lawfulness of acquiring nuclear weapons if it is necessary to deter enemies who already have nuclear weapons. Hashmi supports the position of nuclear pacifists, arguing that Muslims have to act on the basis of Islamic ethics, so they cannot contemplate any use of weapons of mass destruction, and, if they are not usable, their development for deterrence cannot be justified morally, economically or militarily.

The Islamic perspective on the nuclear bomb also has been evaluated by Rolf Mowatt Larssen (2011) in his work *Islam and the Bomb: Religious Justification For and Against Nuclear Weapons*. He analyzes the views of different Islamic scholars on the WMD, especially nuclear weapons, by taking al-Qaeda and Iran as two case studies. After explaining al-Qaeda's pro-nuclear fatwa, he elucidates the position of both Sunni and Shia scholars on the issue. While most scholars agree on the prohibition of initiating a nuclear attack against enemies, even in war, some legitimize the idea of developing nuclear weapons for deterrent purposes and using them for a counter-attack if enemies initiate a nuclear war. More than half of the book is occupied by seven appendices. This part is very useful for acquiring basic knowledge about Islamic texts, scholars, various sects and their ideologies. The fatwa of Tahir-ul-Qadiri (2010) on *Terrorism and Suicide Bombing* also talks about the unlawfulness of indiscriminate killing and the use of WMD.

However, neither Larssen's nor Qadiri's books analyze how these texts impact actual policy-making in the Muslim states. A fatwa is a prescription based on Islamic texts rather than an explanation of the reasons for an existing situation. The edited book by Sohail H. Hashmi and Steven P. Lee (2004) also does not deal with the theoretical aspects of the role of these norms in the nuclear policies of states. Rather than analyzing and classifying the perspectives on WMD within Islam, an analysis of the influence of these perspectives on creating social norms vis-a-vis nuclear weapons and their direct or indirect impact on state's policies is required. The other aspects of religion, apart from its code of conduct, such as religious identity and its role as a base for status and strategic culture and their impacts on nuclear decision-making, are also missing from the existing literature and should be investigated. The present study intends to fill this lacuna in the literature, and provides a new approach to understanding the role of religions in modern states' security-related policy making. It proposes a new perspective on nuclear proliferation and non-proliferation and broadens the framework of existing theories to accommodate religion in their analyses.

This study examines various theories on the causes of nuclear weaponization, adding the role of religion within their frameworks. It also explores religion's direct impact on nuclear policymakers and the indirect influence in nuclear policy by shaping social norms, strategic culture, and public opinion in countries. Thus, the meaning of "role of religion in nuclear decision-making" is taken to include impacts of religious norms, religious identity, religious organizations, and texts. Some of these aspects, such as religious identity, cannot, however, be separated from the secular concept of identity and nationalism. It seeks to investigate the puzzle that if a religion does have an impact, then why does the same religion influence states in different directions? This study considers religion as an independent variable and nuclear policy as a dependent variable. It evaluates the cases of Iran and Pakistan in order to substantiate the arguments and analyses of the perspective of Islam, the state religion of these countries, on nuclear weapons. These cases help understand how the same religion can become either a motivating or de-motivating factor in nuclear policy.

In addition to the introduction and conclusion, this book has four chapters. The first chapter examines existing theoretical explanations of the motivations behind nuclear weapon policies of states. The second chapter, "Role of Religion in Security Policies of States", analyzes the position of religion in the existing literature on international relations and security studies and the possibility of accommodating religion within the existing theoretical frameworks. This chapter also attempts to detail the views of Islam on nuclear weapons. The third chapter studies the case of Iran, describing the historical development of its nuclear program, evaluates the explanations of different theories regarding the possibility of weaponization, and delineates the role of religion in Iran's nuclear decision-making. The last chapter explains Pakistan's policy on nuclear weaponization through the prisms of different theories and evaluates the role of religion as a factor in that policy.

Chapter I

Theoretical Explanation of Nuclearization and Non-Nuclearization

Kenneth Waltz (1954) explains war based on three levels of analysis: structural, state and individual. Theories of nuclear proliferation and non-proliferation can also be examined at these levels. Conventional wisdom on nuclear policy is dominated by structural level 'security-threat' oriented theory of realism. Similarly, constructivism provides another structural level explanation and recognizes international norms as an important factor in shaping state nuclear policies. Theories based on international regimes and access to technology also provide structural-level explanations for nuclear weapons. State-level theories are mainly focused on variables such as economic policies of states, bureaucratic interests, the structure of government and desire for prestige. Psychological motivations and nationalist feelings of leaders and their power in mythmaking are the individual level variables for nuclear policies.

This book looks at these theories and examines their strength and weakness to examine the impact of religion on nuclear policies. Religious norms, which are rooted in the domestic society of a state, are different from international norms and taboos. They are not subject to the criticism that powerful states create norms for maintaining their control at the international level. Religious norms are also significant at the international level. They also influence at the individual level since they shape the psychology of state leaders though this may vary from person to person. So, a theory that considers religion can explain the nuclear behaviour of states, especially nuclear restraint, which structural realism fails to explain.

Structural Level Analyses

Theories that focus on structural-level factors such as security threats, the international regime and international norms are considered as "structural level analyses".

Security Based Explanation

Bradley. A. Thayer (1995: 486) argued that "security is the only necessary and sufficient cause of nuclear proliferation". The theory is based on basic realist assumptions: 1) The international system is anarchic, and there is no central authority to punish aggressor, 2) Great powers possess offensive military capabilities, so they are potentially dangerous and fear each other, 3) Since the intentions of other powers are uncertain, states balance against offensive capabilities of potential rival powers, 4) Survival is the primary goal of every state and 5) States increase their power as a way of self-help (Mearsheimer 2001: 30, Thayer 1995).

States increase their power and balance against major powers through an alliance with other states (external balancing) or by developing their own military capability (internal balancing). Since the alliances are just "temporary marriages" and today's friends may be tomorrow's foe, if they have enough resources and technological capability, states prefer internal balancing. Jacques E. C. Hymans (2010: 456) says that, according to realism, the development of an indigenous nuclear weapon is necessary for two reasons: 1) Lack of credibility of extended deterrence because states cannot expect the US to trade New York for Tokyo or Berlin; 2), Even if these nuclear guarantees are credible, they cannot be depended upon for the long-term security because of the realist principle that "today's ally may be tomorrow's enemy".

Realism provides various military motives for nuclearization of states: The desire to achieve superiority over other states, to prevent rival states from achieving military superiority, to obtain an effective deterrent against the conventional and nuclear threats of enemies, and to obtain the "great equalizer" to the conventional superiority of enemies (William Epstein 1977). The development of nuclear technology by rivals creates security threats to other states and triggers their decision to follow suit for deterrence. Thus proliferation becomes a "strategic chain reaction". Realism explains the historical proliferation of nuclear weapons based on this "proliferation begets proliferation" logic and predicts that the number of Nuclear Weapon States will increase in future, and non-proliferation regimes will be ineffective in preventing it.

Kenneth Waltz (1981) identifies seven possible motives for developing nuclear weapons: 1) Great powers counter rival states usually by imitating their weapons. 2) Lack of confidence in the effectiveness of alliances for retaliating against the attack of nuclear power. 3) States without nuclear allies may respond by obtaining nuclear weapons if their adversaries have them. 4) As a balance against the current or prospective conventional superiority of adversaries. 5) Based on calculations suggesting that nuclear weapons are more economically affordable than conventional weapons. 6) For offensive purposes, but Waltz rejects this motivation. 7) International standing and prestige may be the reason for nuclear development.

These security-based assumptions lead realism to the pessimistic prediction of the gradual proliferation of nuclear weapons. In the 1960s, the US government estimated fifteen to twenty-five new Nuclear Weapon States by the end of 1970s. The famous prediction of John. F. Kennedy indicates this pessimistic view of the US. In the 1970s, about thirty-five new nuclear weapons were projected by the end of 1980s. And again, after the end of the Cold War, rapid nuclear proliferation was forecast, and the absence of such proliferation during the Cold War was explained by constraints imposed by the superpowers (Hymans 2006:3 William J. Long & Suzette R. Grillot 2000: 24).

None of these predictions materialized. This showed the limitations of security-based realism in explaining proliferation and non-proliferation. Realism could not predict the conditions under which states develop an indigenous nuclear weapon for security, rather than depending upon the extended nuclear deterrence of great powers or strengthened conventional weapons. Furthermore, since realism is focused on structural variables and ignores domestic variables and the psychology or ideology of leaders, it could not explain why do two leaders in one state or different states facing similar structural conditions might adopt different policies. Given that many countries have the technological capability to develop nuclear weapons, realism faces difficulties in explaining the reason behind the non-development of nuclear weapons by such states. Why Britain and France and why not Germany? Why Israel and Pakistan and not Iran and Japan and why India in 1998 and not in 1964?

T. V. Paul (2000) criticizes the hegemonic stability theory of realism which suggests that it is the presence of a hegemonic power which halted rapid spread of nuclear weapons by constraining alliance partners through benign policies and non-allied states through coercion. Paul points out that US protection did not prevent some of its allied states from developing their own nuclear weapons. Indeed, most of the proliferation before 1974 was among US-allied states. Even though they are not directly under a superpower nuclear umbrella, the alliance relationship of Israel and Pakistan with the US and India's relationship with the USSR did not prevent these states from acquiring nuclear weapons. Furthermore, supply-oriented proliferation theories suggest that the availability of nuclear technology through alliances or under international regimes for peaceful purposes creates an incentive to develop a nuclear weapon and causes proliferation. Another criticism against hegemonic stability theory is that most of the states signing the Non-Proliferation Treaty (NPT) had not felt hegemonic pressure as the theory would expect. They were also not offered any security guarantee against nuclear attack by another state (Paul 2000: 8).

Paul (2000) suggests "prudential realism" as an alternative to "hard realism". According to this theory, security is not a necessary outcome of nuclearization. In some situations, it may decrease the security of states by creating new conflict situations in the region and causing unnecessary arms racing with other states. Therefore, the decisions of states whether or not to nuclearize depends upon the existing conflict situation and economic interdependence of the region. In short, states belonging to high conflict zones are most likely to nuclearize, and states belonging to moderate and low conflict zones have fewer reasons to nuclearize even though both may have the technological capacity.

"Prudential realism", a combination realism and liberalism, is also not sufficient in explaining the nuclear behaviour of many states. For example, Paul himself has identified the Middle East as a zone of high conflict and minimal economic and trade relations and interdependence (Paul 2000: 10). No other country, however, has developed nuclear weapon for balancing against Israel, and most are signatories to the NPT without a security guarantee from any big nuclear power. This suggests the existence of other constraining variables in the nuclear weaponization of these countries. In short, different realism forms are insufficient to explain proliferation and non-proliferation among states. Realism fails to acknowledge the constraints states are subject to in their path to nuclear weaponization.

International-Regime-Oriented Approaches

The International regime is usually defined as "principles, norms, rules and decision-making procedures around which actor expectations converge in a given issue-area" (Krasner 1983:1). Based on this definition, the international non-proliferation regime can be described by analyzing *principles;* such as nuclear energy can be used without the proliferation of nuclear weapons, *norms;* such as the responsibility of Nuclear Weapon States (NWS) for disarming themselves, *rules;* such as international safeguards and *procedures;* such as NPT Review Conferences.

The role of the international regime in proliferation and non-proliferation decisions has been analyzed from various perspectives. As Thayer (1995: 498) argued that, according to the realist perspective, the NPT and associated regimes cannot prevent nuclear proliferation:

> "NPT and other non-proliferation regimes cannot stop these states from acquiring nuclear weapons; nor should it be expected that the regimes could prevent proliferation. This is because the regime focuses only on the supply side of the proliferation problem......it does not address the cause of proliferation, insecurity of states."

Scholars who argue the importance of international regimes identify, however, two types of effects: regulatory and constitutive. While the institutionalization and enforcement aspects of regime play a regulative role, definition and categorization contribute to the identity and behaviour of states. Institutionalization standardizes behaviour and provides a platform for discussion and dispute settlement. Categorization classifies states, for example, into Nuclear Weapon States (NWS) and Non-Nuclear Weapon States. When the interests of states are determined by an identity established through definition and categorization, institutionalization and enforcement reinforce the categorization (Sasikumar 2006: 57). However, regulative and constitutive effects can be differentiated from each other. Regulative aspects affect the interest calculations of states directly through an explicit process of international institutions, while constitutive impacts are indirect and hard to study because they have not been explicitly documented (Sasikumar 2006: 12).

The analysis of regulative and constitutive effects helps make intelligible the supply and demand factors behind nuclear proliferation decisions. Most studies on the role of regimes focus on regulative aspects and technology supply. They expect that nuclear proliferation can be arrested by creating constraints over technological availability and by ensuring effective safeguards and implementation of rules through "sticks and carrots". Supply-side theories consider regulation over the transfer of technologies, sharing of information about suspected nuclear aspirants and sanctions will halt the spread of nuclear weapons.

A mere focus on the regulatory aspects of regimes leads to various criticisms. A major criticism is that nuclear regimes do not address the motivations behind a state's desire for nuclear weapons. If a state really wants to make nuclear bomb, regimes cannot stop it. Thayer (1995: 500-506) identifies many problems in regime theory. He says, "regime can at best only delay, not stop, the nuclear weapon program". Even such a delay may not be possible due to the nuclear assistance of great powers to emerging nuclear powers by promoting their national interests over the non-proliferation interests of the regime. For example, assistance from China to Pakistan and from France to Israel were crucial for their nuclear development.

Since technology and materials for nuclear development are available in black-markets and through secret trade with some states, regimes may not significantly increase the costs of nuclear-weapon programs. The NPT has been criticized (J. Erickson and C. Way 2011) for allowing international trade in peaceful nuclear technology. It is argued that the availability of nuclear technologies for energy purposes enables states to develop nuclear weapons without nuclear testing and thereby causes "opaque proliferation". Jacques E. C. Hymans (2006) criticizes regime-oriented theories by asking whether regimes have halted any state that otherwise would have developed a weapon or they simply reinforced the already existing non-proliferation commitment of states. Hymans argues that even before the emergence of non-proliferation regimes in the 1970s, there was a gap between potential nuclear powers and actual nuclear-weapon states and offered the graph shown in figure - 1.

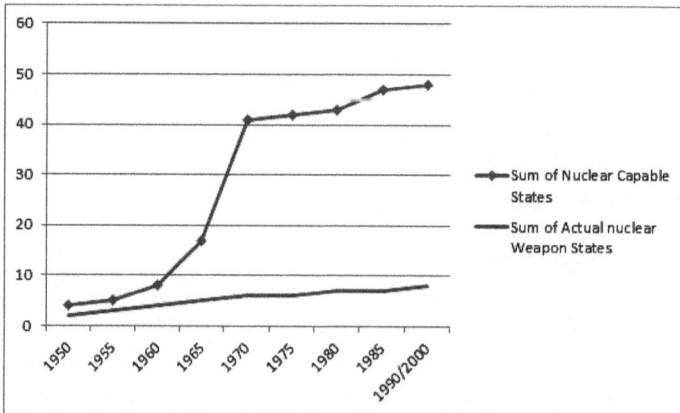

Figure-I

Nuclear Capable States and Actual Nuclear Weapon States
(Source: Hymans 2010: 457 and Hymans 2010: 4)

Figure 1 shows that the absence of proliferation is not because of the lack of technological capacity but because most states prefer to be non-nuclear for other reasons. Regime theory fails to give a convincing explanation for some states agreeing with the rules and norms of a regime while others do not. It is not clear whether the regulative mechanism of non-proliferation regime prevents signatories from proliferating or that states do sign these treaties because, for other reasons, they already have no plan to go nuclear.

Most criticisms of regime-oriented explanations of non-proliferation question the direct impact of their regulative aspects, i.e., institutionalization and implementation. The indirect impacts of constitutive aspects, such as definition and categorization, can overcome many of these criticisms because they consider the demand side of proliferation that motivates or constrains nuclear weaponization. However, the role of domestic and individual elements in policy making and a state's response to international regimes must also be considered. The constitutive aspects of the regime have no direct effects on nuclear policies. However, the decisions of state leaders, the definition of the basis for national prestige and domestic norms are influenced by international regimes. The constitutive aspect of the regime is more related with constructivism than liberal institutionalism. It deals with the overlapping aspects of international regimes and international norms. However, just as with mainstream constructivist theories focusing on international norms, international regimes theories also have limitations to counter-realist criticism that argues that international norms are products of great power politics.

International Norms Oriented Approach

International idealists explain proliferation and non-proliferation based on norms at the international level. The state, as a member of international society, is influenced by these norms. When a state acts in international society, its actions are affected by the common expectations of other states. States want to keep a "good name" among other states and strive to increase their relative status. States expect that a status of 'good state' will be useful for alliance formation and as soft power. Nuclear policies of states also depend upon the normative aspects of nuclear weapons in international society. Nuclear norms are dynamic and have changed over time from the nuclear weapon being a symbol of prestige and legitimacy to one of illegitimacy and of being a rogue state. Therefore, both proliferation and non-proliferation can be explained by an international norms-oriented approach. States seek nuclear weapons when possession is considered as a basis for prestige and membership in the great-power club.

11

Scott Sagan (1996: 73) says:

> "According to this perspective, state behaviour is determined not by leaders' cold calculations about the national security interests or their parochial bureaucratic interests, but rather by deeper norms and shared beliefs about what actions are legitimate and appropriate in international relations".

At the same time, Nina Tannenwald (2007) explains non-use of a nuclear weapon on the basis of a "nuclear taboo" that developed at domestic level in the US and has diffused to other countries and to the international level. She identifies three effects of norms: regulative, constitutive and permissive. International nuclear norms regulate the behaviour of states and constitute identities such as "civilized states". She argues that permissive nuclear norms create a category of Weapons of Mass Destruction and permit other forms of weapons and violence. According to Ronald L. Jepperson, Alexander Wendt, and Peter J. Katzenstein (1996), "norms shape the national security interests or (directly) the security policies of states". Norms define national identities in particular contexts (constitutive effects) and prescribe or proscribe the expected behavior of these actors (regulative effects).

Most constructivist theorists considered international norms as an effective constraint on the actions of member states. This approach has limitation to separate the interests of great powers in the formation of these norms. During the interwar periods, idealism was criticized by realism on its ignorance of national interests of great powers in international norms and morality. E.H Carr (1939: 79) in his celebrated book "The Twenty Years' Crisis" said:

> "'Theories of social morality are always the product of a dominant group which identifies itself with the community as a whole, and which possesses facilities denied to subordinate groups or individuals for imposing its view of life on the community."

The international norms on nuclear weapons have been criticized as a product of the interest of existing Nuclear Weapon States (NWS) in keeping their dominance over this ultimate weapon. Nina Tannenwald responds to this criticism by arguing that the emergence of a taboo on the use of nuclear weapons in the first decades of the Cold War was not in the interest of nuclear powers especially the US, which depended on nuclear weapons to balance the superiority of the USSR in conventional weapons. Furthermore, grassroots anti-nuclear movements, the United Nations and non-nuclear weapons states were major players in creating international norms (Tannenwald 2007:60-61). However, it cannot be ignored that the interests of the US play a role in present anti-proliferation activities at international level.

Another difficulty of the international norms-oriented approach is understanding the role of domestic society in creating these norms. Many countries, especially third-world countries, took a stand against nuclear weapons in normative and moral terms even before the emergence of strong international norms. So, Tannenwald's explanation of the nuclear taboo as emerging from the domestic level in the US and diffusing to other countries is not enough to explain the development of norms in India and other countries. Therefore, this study, while accepting the importance of norms in nuclear policies, questions the supremacy of international norms over domestic norms and considers the relevance of domestic level analysis in identifying the origin of nuclear-related norms.

Domestic Level Analyses

Second-level explanations focus on domestic variables such as the economic policies of states, types of regimes, public opinion, domestic norms, aspirations for prestige and bureaucratic interests. Domestic-level explanations are useful to understand the variation in responses of states to structural level factors. They help answer the question as to why do some states develop nuclear weapons and others do not when they face similar structural conditions such as external security threats. However, domestic-level explanations are not unanimous in pointing to a single variable.

Economic Policy-Oriented Approach

Etel Sollingen (1994) explained the nuclear behaviour of states and shifts in policies over the years by focusing on domestic economic coalitions. She identified two coalitions: those advocating economic liberalization and those opposing it. Sollingen connects nuclear policies of states with this economic orientation. The first group, which advocates for economic liberalization, is internationalist in its worldview. They accept nuclear weaponization to maximize benefits from the international market when it does not endanger the state. These coalitions "rely extensively on the global economy and the political support of major powers within regimes" and choose the non-proliferation path to smooth relations with them. The second group, an "inward-looking" coalition, opposes economic liberalization by focusing on the distributional consequences of plans imposed by global financial institutions. They prefer to make their states self-sufficient by reducing dependence on international institutions and great powers. These coalitions consider the development of nuclear weapons as a way to autonomy.

This theory, proposed by Sollingen, is helpful to understand the shift in nuclear policies of a state over time and variation in policies of different states even though the structural situations are similar. However, a weakness of this theory is that it excludes the first-generation Nuclear Weapon States from its analysis and focuses only on second-generation nuclear "fence-sitters". This theory finds it difficult to explain the motivating factors behind the nuclear development of nearly half of the total nuclear weapon states. The cases provided by Sollingen also stand to refute her theory. For example, Iran pursued the development of a nuclear weapon under the Shah (Saira Khan 2010). However, after the Islamic revolution, the new regime denounced nuclear weapons (Sohail H. Hashmi 2004: 342) and followed an inward-looking policy. The case of India also does not provide sufficient support for Sollingen's argument. Even though the Bharatiya Janata Party (BJP) expresses higher level nationalist attitude compared to Indian National Congress (INC), it is difficult to argue that the economic policies of BJP-led government in 1998 were more inward-looking than those of the Congress-led governments of previous decades under the leadership of Jawaharlal Nehru and Indira Gandhi. Even today, the BJP-led government liberalizes the economy while claiming to express a high level of nationalism. Jawaharlal Nehru and Indira Gandhi were against nuclear weapons even though they followed inward-looking nationalist economic policy. The case of Iran shows that states may follow non-proliferation path even under nationalist and inward-looking governments and the case of India shows that the inward-looking leaders may follow the non-proliferation path while an internationalist coalition may develop nuclear weapons.

William Epstein (1977) discusses various economic motivations of states in their nuclear decision: 1) States consider nuclear energy as a cheaper source of energy and useful for improving the economy and living standard of the state. 2) Some – mainly third-world states – pursue peaceful nuclear explosives. 3) "Potential military spinoff benefits" motivate some countries to build up peaceful nuclear industries. 4) Some states believe that possessing nuclear energy helps them reduce the gap between themselves and rich countries. 5) Establishment of a domestic nuclear power industry reduces the cost of nuclear weapon development at any future time. States also consider nuclear weapons cheaper compared to conventional weapons. 6) It is of major economic interest, mainly for former colonies, to avoid dependence on super-powers in both political and economic terms. 7) Some third world states consider nuclear weapon a source of leverage in their demand for New International Economic Order. However, Epstein's explanations, which were written after India's 1974 Peaceful Nuclear Explosion (PNE), focused mainly on the reasons for the development of nuclear power plants and peaceful nuclear tests. If states developed their nuclear programs because of these motivations, the number of nuclear states would be much larger. Thus, economic theory is not enough to understand "why states do and do not go nuclear".

Domestic-Regime Oriented Approach

Domestic political system is one of the dominant variables in the analysis of nuclear policies (Long and Grillot 2000: 26). The approach based on political system is developed from Kantian 'democratic peace theory'. The democratic peace theory proposes that democracies do not go to war against each other because of two reasons. First, institutional structure and dynamics ensure the check and balance system and scope for public debate on important issues. This theory presumes that people usually prefer peace over war. Second, democracy creates a culture of accommodation, cooperation and peaceful dispute settlement rather than conflicts and use of force. Relations between democracies are shaped by this norm. This thesis is the essential pillar of liberal and neoliberal theories.

However, as Etel Sollingen says, there was no systemic approach to extend this theory into the field of nuclear proliferation (Sollingen 1994: 131). The democratic peace theory and trust of one democracy in another may have both positive and negative influences on nuclear development. Nuclear weaponization may be constrained in democratic countries due to the influence of public opinion, accountability of government, check and balance system and the culture of a peaceful settlement. It raises questions about the relevance of public opinion in making foreign policy. H. Morganthau (1948) has warned about the danger involved in preferring unsound public opinion while deciding on foreign policy. However, according to Long and Grillot, "the internalization of democratic norms could constrain a state's nuclear ambition, or cause a state to relinquish a nuclear arsenal it already possesses because democratic norms make it unacceptable for democracies to deter other states with one of the most dangerous and violent forms of weapons" (Long and Grillot 2000: 26).

Nonetheless, at the empirical level, the fact that majority of Nuclear Weapon States are democratic questions the optimist view of democratic peace theory. Karsten Frey says that six out of eight nuclear-weapon states and twenty-eight out of thirty nuclear-capable states are democracies (Frey 2006: 15). This reality of nuclear trend among democracies leads us to different reasons. First, as an impact on democratic peace theory, one nuclear-capable democracy may transfer the technology to another democracy, believing that it may not threaten them. Second, as supply side oriented theories and technological determinism argue, the availability of technology may motivate states to develop nuclear weapons.

Just like democracy does not constrain states from nuclear technology development; there is no scope for optimism on democracies about the use of a nuclear weapon or the threat of using it. The only country that has used nuclear weapon ever in the battlefield is a democracy, i.e., the USA. When states considered as non-democracies like China and USSR (during the Cold War) followed no-first-use policy and refrained from use or threat of nuclear weapons against non-nuclear states, democracies like Britain, France, Israel and the US did not have such no-first-use policy and they also threatened the use of nuclear weapon many times even against non-Nuclear Weapon States (See: Paul 2009:19-20). In short, democratic peace theory has a limitation in explaining the nuclear behavior of states.

Glenn Chafetz (1993) provides a different explanation for the relationship between democracy and nuclear weapons. He divides the world into two categories: The Core States and the Periphery States. Chafetz argues that since the domestic political system of core states is rooted in neoliberal democracy, they share common values and norms of international cooperation rather than conflicts and arms racing. However, periphery states do not have such domestic democratic norms and hence, they do not have shared values at the international level for cooperation (See: White 1996: 49). This argument has many limitations. First of all, international cooperation and institutions of third world countries such as NAM and G-77 refute Chafetz's argument that periphery states do not have shared values of cooperation at the international level. Another major drawback of the theory is that majority of the nuclear states belong to the 'core' region. This theory faces difficulty in responding to empirical realities that are exactly opposite to its expectation, i.e., nuclear proliferation among democratic states belong to the core region and non-proliferation and anti-nuclear movement originated from the peripheral states through their shared values and norms.

Bureaucratic Structure, Organizational Level or Decision-Making Process Oriented Approaches

Unlike structural-oriented theories, which ignored the domestic bases of nuclear policies and economic policy and regime type oriented theories, which considered domestic aspects of states as a single unit without considering the internal diversities, bureaucratic structure or organizational level oriented theories consider internal debates and domestic discussion in the process of nuclear decision making. These theories explain how parochial bureaucratic and political interests shape the attitude of decision-makers. The outcome of the internal debates reflects the interests of dominant group at the domestic level. That outcome is seen as the opinion of the state. So the nuclear policy of states reflects the interests of dominant groups at the domestic level. Some celebrated works such as Organizational Theory of Scott Sagan (1995) and his second model out of three model approaches (Sagan 1996), and bureaucratic politics model of Allison deal with the domestic differences over nuclear decision-making.

William Epstein says that in every country there are competing opinions, and different groups try to make their view prevail. On one side, the military-industrial complex and scientist and bureaucrats associated with nuclear program will promote nuclear programs. On the other, peace groups and scientists in the academic institutions will oppose the nuclearization. The outcome depends upon the strength of each group in their bargaining (Epstein 1977: 25). S. Sagan (1996) presented his second model of explanation for nuclear proliferation, focusing on "domestic actors who encourage or discourage governments from pursuing the bomb". He identified three kinds of actors: those works in nuclear energy establishment of the state, important units within the military and political leaders. Sagan explains the nuclear policy of India based on this 'second model'. He argues that the Indian response to Chinese nuclear test of 1965 was not united, but the test produced "a prolonged bureaucratic battle" (Sagan 1996: 66). According to Sagan, this bureaucratic bargain is to be considered in order to solve the puzzle of India's development of nuclear weapon, i.e., the absence of nuclear test after its defeat to China in 1961 or after nuclear test by China in 1964, but it tested nuclear weapon after victory over Pakistan in 1974. This indicates that external threat and response to it depends upon how different groups within the domestic sphere interpret the situation according to their interest and logic. The domestic political situation also shapes the interest of these groups. Some groups argue for the development of nuclear weapons to improve their influence at domestic level.

Graham Allison (1971) also gave importance to the decision-making process and domestic differences among bureaucrats and "self-interested bargain among intrastate actors". In his words "different groups pulling in different directions produce a result .....distinct from what any person or groups intended....in both cases, power and skill of proponents and opponents of the action in question" (Allison 1971:145, Quoted in Thayer 1995: 475). Both Allison and Sagan question the realist assumption that the state is a unitary actor.

The focus of this theory is on the interests of the groups as a whole, not the interests of each individual. Because, according to Sagan and Allison, interests and rationality are bounded by the structure of the organization. Sagan says

> "Large organizations function within a severely "bounded" or limited form of rationality: they have inherent limits on calculations and use of a simplifying mechanism to understand and respond to uncertainty in the outside world" (Sagan 1995: 51).

According to Allison, individuals are retrained by the bureaucratic structure and therefore, their behaviour can be predicted by their role in the bureaucracy. In Allison's own words "where you stand depends on where you sit" (Allison 1971).

However, the bureaucratic or organizational models are not immune to criticism. Bradley A. Thayer questions the validity of bureaucratic-oriented theory on the basis of counter-factual logic. He asks, if Homi Bhabha had never existed, would India have not developed the nuclear weapons? (Thayer 1995: 478). However, this criticism ignores the basic concept of the bureaucratic theory that it is the bureaucratic structure, not an individual, which is the pushing factor behind nuclear weapons. The nuclear pull from Bhabha is not as an individual but the position and structure to which he belongs. Whoever works in that position will take the same decision. However, Tanya Ogilvie-White puts forward a more powerful criticism of the bureaucratic or organizational model theory. She argues that the explanatory power of organizational theory is limited due to following reasons. First is the agent-structural problem. By focusing on the structural explanation of behavior, organisational theory ignores the role of individuals and their influence in nuclear decision-making. Second, even though the theory says that organizations have influence, it does not explain which organization is more influential and why? Third, by overemphasizing the influence of organizational culture, the theory leads to "unnecessary deterministic and pessimistic outlook" for the non-proliferation attempt. The theory ignores the fact that individuals and organizations can learn from past experiences and new information may lead to changes in the organizational culture (White 1996: 51). The bureaucratic or organizational theory can be criticized as they are too abstract and cannot explain non-proliferation and constraints over bureaucrats in their path towards nuclear weapons as per their aspiration. The theories based on the role of domestic norms can overcome these criticisms.

Domestic Norms and Culture Oriented Theories

The focus of constructivism is mainly on the international norm and its effects on policies of states. However, domestic norms and culture have significant influence in constituting and regulating the worldviews of leaders and legitimizing their policies. The advantage of focusing on the domestic level is that it can cover norms of Third World States which are different from Western developed states. Even the norms which are similar to western norms are not necessarily diffused from western society. They can be rooted in the traditions of the Third World States themselves. Many norms of Third World Countries have not developed into international norms. As the theory of 'particularism' says, norms are culturally defined and norms of one culture may be different from another. Therefore, focusing on domestic norms is a better way to understand the nuclear behaviour of Asian States than restricting the analysis to international norms.

Domestic norms and culture are significant sources of strategic culture of states. For example, Thomas U. Berger (1996) explains how the anti-militaristic norms of Germany and Japan made it difficult for them to follow more assertive national security policies after Cold War. Since culture refers to "collection of value preferences" or "set of standards that define the existence, operation and mutual relations of social actors", an explanation based on the culture can cover ends, means and strategies of states (Jeannie L. Johnson et al. 2009: 8). The explanations of mainstream theories such as realism are based on rationality. They presume that states are rational actors and predict the policies of states according to this rationality. However, Jeannie L. Johnson et al. (2009: 6) quote Valerie Hudson arguing that "rationality itself may mean different things to different cultures". It is necessary to understand the cultural and normative context to explain "irrational" behaviour of other states. Ronald L. Jepperson et al. (1996) identify three different effects of the cultural environment. First, the environment affects the behaviour of actors. Second, it affects the contingent properties of actors such as identities, interests and capabilities. Third, it may affect the existence of actors altogether. However, this effect of culture or norms on actors is not unidirectional. The leaders have a significant role as "norm entrepreneurs" in choosing one norm out of many competing norms and in mobilizing people according to their interpretation of norms.

Norms and culture have crucial impacts on the nuclear decision-making of states. They set the standards for the acquisition, proliferation, and use of nuclear weapons. The leaders of states use norms to legitimize their nuclear policies within domestic and international communities. Norms also facilitate "mythmaking" of leaders regarding nuclear weapons. Kerry M. Kartchner (2009) explains three links between strategic culture and WMD-related decisions. First, acquiring, proliferating or use of WMD (Weapons of Mass Destruction) deemed rational within that culture. Second, acquiring, proliferating or use of WMD are perceived by holders or keepers of strategic culture as enabling the state to achieve culturally endorsed outcome such as a means to defend its perceived enemies or increase its prestige. Third, "the end and means for achieving the culturally endorsed outcome (acquiring, proliferating or use of WMD) are consistent with the 'repertoire or palette of adaptive responses' deemed appropriate by holders of that strategic culture".

Culture and norms of states can be the motivation behind either nuclear weaponization or non-weaponization. For example, Iranian public statement refers to its culture and religious prohibition of weapons of mass destruction to reject nuclear weaponization. Some cultures, which allow acquiring of nuclear weapons, prohibit its use. The existence of taboo against the use of nuclear weapons is an example. However, these normative preferences may change due to many factors like a challenge to existing belief by external threats. The conflicts between different norms and change in preference of leaders also may lead to shift from one norm to another. For example, if the moral stand of Iran gets challenged by an external threat or preferences of leaders change from peace through non-weaponization to peace through deterrence, the nuclear policies of the state would also change. However, states have to reshape public opinion and norm of non-weaponization into the norm of deterrence and security through nuclear weapons. For this purpose, like what happened in India and Pakistan, holders of national culture use religion and interpret its principles to justify weaponization. Therefore, national norms can be either real cause and independent variable or an instrument and intervening variable in nuclear decision-making. The theory which is based on domestic norms can explain the difference in policies of states even when they face similar structural condition.

Prestige and Status Oriented Approaches

This study considers prestige and status under one theme due to similarities of their impacts in nuclear policies of states. Some scholars have conceptually differentiated between prestige and status. Larson, Paul and Wohlforth (2014: 7) defined status as "collective beliefs about a given state's ranking on valued attributes such wealth, coercive, capabilities, culture, demographic position, sociopolitical organization and diplomatic clout".

They consider status as collective, subjective and relative. They differentiate status from prestige, which is defined as "public recognition of admired achievements or qualities", because status can be referred to ranking on a hierarchy, but not prestige. There is a possibility of first, second or third ranks in status, but such rankings are not suitable to prestige. However, even though there is a possibility of conceptual difference, the impacts of both status and prestige on nuclear policies are the same. Additionally, scholars have cited the same example for proving status and prestige-oriented explanation. For example, the words of Charles de Gaulle, "will France remain France?" has been quoted as an evidence of both prestige and status. Furthermore, Larson, Paul and Wohlforth (2014:12) refer to the works of Barry O' Neill, entitled as "Nuclear Weapons and Pursuit of Prestige" as a work on status-seeking behavior in nuclear policy. It confirms that the distinction between status and prestige is irrelevant in explanation of their impacts on nuclear policy. Both prestige and status can be the motivating factors behind proliferation or non-proliferation.

States look to enhance their prestige in the international sphere either as an end in itself or as a means to other ends. When it is considered as an end in itself, states increase various forms of power to achieve prestige. Prestige, sometimes, becomes a means for other ends like having influence on other states, to increase its soft power and to attract other states to alliance formation. However, even though almost all states look to improve their status before others, the policies for this purpose vary depending on the social conditions and interpretation by leaders of individual states. For example, nuclear weapons were a symbol of prestige at one point of time, and it gradually shifted into a scenario where violators of non-proliferation norms are considered 'Rogue States'. At the same time, viewpoints of states and leaders on the impact of nuclear weaponization on the status of the nation may differ according to the domestic norms of states and interpretation of leaders.

The theory, which considers desire for prestige as the motivation behind nuclearization of states, assumes that "all great powers must have nuclear weapons". William Epstein (1977) points out the different prestige-oriented motivation of states in their nuclear policy. First, states build a nuclear weapon to achieve or maintain great power status. Second, they aim to ensure a seat in the "head table" of international institutions. Third, they look to improve their prestige within their region. Fourth, states, mainly former colonial countries, develop nuclear to redress their inferiority at the international level. Fifth is to save themselves from international level discrimination between Nuclear Weapon States (NWS) and Non-Nuclear Weapon States (NNWS). Sixth is to demonstrate the ability of the state in self-reliance without depending upon superpowers (Epstein 1977: 21-22).

However, there are various sources for national prestige. Sometimes it may include opposite behaviours. For example, in some occasions, holding the territory of other states is a source of prestige while on different occasion invasion to a foreign territory will negatively affect the prestige of states. Barry O' Neill (2002) categorizes the sources of prestige into ten groups. They are military possession and action, moral actions like foreign aid, intellectual, cultural and sports achievements, standing up to an adversary without being defeated, holding foreign territories like colonies, economic power and other internal strength of states, recognition by other states like membership in international organizations, independence and assertiveness in policies, foreign involvement through trade and other ways, and possession of allies.

Bradley A. Thayer (1995: 471) criticizes the prestige-oriented theory arguing, "prestige is neither a necessary nor sufficient cause of proliferation". He puts forward three reasons to support his argument. First, the validity of the theory is flawed, because nuclear weapon is not the sole criteria for great power status. Many states like Germany and Japan are great powers without a nuclear weapon. At the same time, some nuclear weapon states like Pakistan and Israel are not treated as great powers. Second, it is extremely unlikely for a state to spend a huge amount of budget just for the purpose of prestige. Third, empirical evidence does not support the theory. Even the proponents of this theory mention only France, India and UK, not rest of the nuclear states, as examples for developing a nuclear weapon for prestige purpose.

This study argues that prestige is an important motivation for states in their foreign and military policies. As described before, prestige is useful both as means and end. These include the idea of prestige for a state, prestige for leaders within and outside their states or prestige of a common identity or nation which spread across boundaries of various states. For example, the Islamic nationalism in Pakistan motivates leaders to consider their nuclear weapon as a source of prestige to all Islamic countries. Similarly, third world solidarity concept of India may motivate the Indian state to consider its nuclear weapon as prestige for all Third World States. So, the prestige-oriented explanation overlaps all three levels of analysis: individual, domestic and structural. However, since the domestic aspect is dominant, this section is placed under the category of domestic level analysis.

Individual-Level Analyses

The explanation of international politics based on the character of an individual is not new to discipline of International Relation. Daniel L. Byman and Kenneth M. Pollack (2001: 110) argue that at least since Aristotle, scholars have tried to explain politics focusing on individuals. Classical realists like Thucydides, Niccolo Machiavelli and Morgenthau explain international relations based on the nature of human being. The individual level explanation of nuclear policy mainly focuses on the psychology of individual leaders, their nationalist conception, their personal interests to improve their position at domestic level, and the role of individuals in mythmaking.

Psychological Approach to Nuclear Policy

The psychological analysis of nuclear policies identifies that nationalist feeling of leaders influences the nuclear policies of the state. The idea of nationalism may include different types of nationalism such as ethnic, religious and civic nationalism. This nationalist feeling may also go beyond the boundaries of states if the people of different states belong to same identity like religion. The nationalist feeling of the individual varies from leader to leader. Jacques E. C. Hymans (2006) in his work, *The Psychology of Nuclear Proliferation*, investigates the different mode of 'National Identity Conception' (NIC) and explains nuclear policies of states based on it. He defines NIC as:

> "an individual's understanding of the nation's identity – his or her sense of what the nation naturally stands for and of how high it naturally stands, in comparison to others in the international arena" (Ibid: 13).

He classifies the NIC into four categories: Oppositional Nationalist, Sportsmanlike Nationalist, Oppositional Subaltern and Sportsmanlike Subaltern.

Hymans uses the definition of *Routledge Dictionary of Politics* for his further explanation and classification of nationalism. According to this definition, nationalism is "the political belief that some group of people represents a natural community that should live under one political system, be independent of others and, often, has the right to demand equal standing in the world order with others" (David Robertson 2003: 331, quoted in Hymans 2006: 24). Hymans figures out different types of the National Identity Conceptions by analyzing how leaders consider their nation in their relation with other nations. He checks the "solidarity" dimension and the "status" dimension. In the "solidarity" dimension, the attitude towards another state may be "us and them" or "us versus them". The first one is "sportsmanlike NIC" and the second is "oppositional NIC". The oppositional NIC of "us against them'" creates 'fear' among the leaders. The key question in status dimension is whether one state is naturally equal (if not superior) or inferior to another state. Hymans identifies the feeling of equality or superiority as "nationalist" and feeling of inferiority as 'subaltern'. The nationalist feeling creates 'pride' among leaders. The leaders with oppositional nationalist NICs are characterized by a mix of both 'fear' and 'pride'. Such leaders are the individual force behind nuclear weaponization of states. In Hymans' words

> "Oppositional nationalists see their nation as both naturally at odds with an external enemy and as naturally its equal if not its superior. Such a conception tends to generate the emotions of fear and pride – an explosive psychological cocktail. Driven by fear and pride, oppositional nationalists develop a desire for nuclear weapons that goes beyond calculation, to self-expression" (Hymans 2006: 2).

In short, Hymans considers the nature of oppositional nationalist leaders with fear and pride as a major pushing factor behind the nuclear policy of states. However, the difficulty with this approach is measuring the degree of nationalism. Hymans has tried to quantify the degree of nationalism by analyzing the speeches of leaders. Another flaw of NIC oriented theory is that it does not consider the domestic and international situations which shaped the NIC of leaders. Even though the position of leaders is important in the ultimate decision on a nuclear weapon, a top-down approach without considering the social condition of those leaders and the factors behind their mode of NICs is inadequate to explain nuclear policies.

Mythmaking and Nuclear Weapons

Peter R. Lavoy (1993 and 2006) develops another individual level psychological approach of nuclear weaponization by focusing on "Nuclear Mythmaking". Lavoy points out how views of individual affect nuclear policies of states. Unlike Hymans, who focused on NIC of top leader to understand states' behaviour, Lavoy recognizes the existence of competing myths among strategic elites at domestic level. In his words "at any given time and in any given country, multiple strategic myths may coexist and compete with one another" (Lavoy 2006: 435). He focuses on internal debates among those who have different views on the nuclear weapon. The nuclear behaviour of a state in each period depends upon the regime which is prevailing in that period. Homi J. Bhabha is a perfect example for nuclear mythmaker in India. However, he could not succeed to compete with prevailing nuclear mythmaking of Nehru and Gandhi, who stood for universal disarmament. Hassan Abbas (2008) has explained proliferation motivation of A.Q. Khan's network from Pakistan to Iran, Libya and North Korea by using the concept of nuclear mythmaking.

The mythmaking theory has similarity with bureaucratic or organizational oriented theories in its aspects of internal debates and competition. In both cases, the nuclear behavior of a state depends on which group prevails each time. However, there is a major difference between bureaucratic-oriented theory and nuclear mythmaking theory. On the one hand, the individual agency has an important role in mythmaking (Lavoy 1993: 199) while on the other, according to bureaucratic or organizational oriented theories, structure and position of individuals constrain their thoughts. This difference is the main reason for this study to classify bureaucratic or organizational oriented theories into domestic level analysis and mythmaking theory into the individual level analysis.

According to Lavoy, a state likely to develop nuclear weapon when national strategic elites, who support nuclear weaponization

> "(1) emphasize their country's insecurity or its poor international standing; (2) advance this strategy as the best corrective course for these problems; (3) articulate the political, economic, and technical feasibility of acquiring nuclear technology/weapons; (4) successfully associate these beliefs and arguments (nuclear myths) with existing cultural norms and political priorities; and finally (5) convince senior decision makers to accept and act on these views" (Lavoy 2006: 435).

However, since the multiple myths are present and competing each other, each mythmaker cannot succeed in achieving their goal. The success of a myth depends on three factors: first, compatibility of the myth with prevailing cultural norms and political priorities; second, the ability of mythmaker to popularize his myths and convince national leaders; third, integration of strategic myths into organizational identities and missions of institutional actors (Ibid. 435). Karsten Frey (2006) differentiates between 'nuclear myth' and 'nuclear taboo' and explains them as opposite terms. He considers nuclear weaponization as a result of nuclear myth and non-proliferation as a result of nuclear taboo. However, the success of either myth or taboo, according to Frey, or one of many myths, according to Lavoy, depends upon various social factors.

According to both Frey and Lavoy, nuclear experts have a key role in nuclear policy-making. Their analyses based on political, strategic, economic and normative reasons, can influence top-level leaders and public opinion. However, as Lavoy points out, social and religious norms and identities are important factors in the nuclear mythmaking and popularizing process. Hassan Abbas (2008) also considers religious identity as a major factor behind nuclear mythmaking of A. Q. Khan's network. It refers to various aspects of the impacts of religious norms and identity on 'nuclear taboo', 'nuclear myths' and 'nuclear nationalism'.

Chapter II

Role of Religion in Security Policies of the States

This chapter deals with theoretical understandings of the role of religion in nuclear policies of states. This chapter attempts to investigate the following questions: What is a religion? What is the role of religion in the policies of states and how did various historical events influence the relation between states and religion? How do various theories of International Relations and Security Studies accommodate religion? Finally, this chapter summarizes different views within Islam on nuclear weaponization and analyzes whether Islam motivates or demotivates nuclear weaponization.

What is Religion?

Scholars of International Relations (IR) or social sciences have not reached a consensus on the definition of religion. David Laitin pointed out that "no consensus exists as to what religion is" (Quoted in Michael C. Desch 2004). William T. Cavanaugh (2009: 3-5 and 57-60), in his *Myth of Religious Violence,* argues that there is no transhistorical and transcultural concept of religion, and it is impossible to separate religion from economic and political motives. Scholars like Bruce Lincoln argue that "no universal definition of religion is possible since all such definitions are products of specific historical and cultural context" (Quoted in Luke M. Herrington and Alasdair McKay 2015: 10-11). Brian S. Turner (1991) (quoted in Jonathan Fox and Shmuel Sandler (2004: 176) refers to a dozen and contradictory definitions of religion in social sciences. He says that it is hard to reach a definition which covers all religions of the world. One of the main reasons for this lack of unanimity is the absence of uniformity among world religions. The western concept of religion with belief in some gods or divine entities is not suitable for 'Eastern religions' like Buddhism.

Early social scientists, as Herrington and McKay (2015: 10-11) pointed out, considered only Christianity, Islam, Judaism and Paganism as religions and observed that 'savages have no religion.' They treated Islam using the western idea of Christianity and argued that Islam is pre-modern since it does not separate religion and state. Talal Asad (1993:1) has warned against the normalizing concept of religion when translating Islamic tradition.

By considering the difficulty in providing a universal definition of religion, scholars like Jonathan Fox and Shmuel Sandler (2004) focus on what are the influences of religion rather than relying on a specific concept which defines it. To understand the nature of religion and to overcome the difficulties in defining it, William P. Alston identifies seven elements of religion. They are 1) a belief in supernatural; 2) the ability to communicate with supernatural; 3) a belief in some form of transcendent reality; 4) a distinction between the profane and the sacred; 5) a worldview articulating the human role in relation to the world; 6) code of conduct, and 7) a temporal community bound by adherence to the preceding elements. Even though no religion may be characterized by all of these seven elements, most of these are features of many religions (Luke M. Herrington and Alasdair McKay 2015: 10-11)

Cavanaugh (2013: 56-67) identifies four approaches to religion in International Relations literature. First two of these approaches consider religion as *sui generis*. According to these, religion is regarded as something distinct from other elements of human culture which are labeled as "secular". The first approach argues that International Relations is a secular social science, and therefore, it should avoid religion. Realist approach is an example of it. The second approach argues that even though religion is a separate entity, it is a significant factor in international relations. So, it is an important area for study. Third and fourth approaches do not consider religion as *sui generis*. They believe that religion is not a distinct factor from other aspects of life. The third approach regards religion as reducible to other factors. For example, a Marxist may consider religion as a superstructure and secondary effect of a basic economic factor. The fourth approach is convinced by the evidence that the "distinction between 'secular' and 'religious' is not a trans-historical and transcultural aspect of human life, but it is an invention of modern West." According to this approach secular-religious division is a construction of the West and exported to rest of the world through colonization.

Recognizing the difficulties in reaching a universally accepted definition and limitation of such a definition, this research follows the definition of Johnston (1994) as "an institutional framework within which specific theological doctrines and practices are advocated and pursued, usually among a community of like-minded believers." This definition includes belief system, practices, identity, and organizational aspects of religion. So, religion can be seen as an ideology, texts, identity, norms, local practices, organization or way of life. So, the meaning of "role of religion in nuclear decision-making" includes impacts of religious norms, religious identity, religious organization, and religious texts. However, some of these aspects, such as religious identity, cannot be separated from the secular concept of nationalism and status. For example, the influence of religious nationalist feeling on nuclear policies or seeking of status for religious identity through nuclear weapons cannot be seen as purely religious one. At the same time, religion can assist other causes to reach their destination. Therefore, the influence of religion on state policies can be either an independent variable or an intervening variable.

Religion as a Neglected Factor in International Relations

The origin of social sciences, especially International Relations, was in the context of secularization in society and academic disciplines. Religion was separated from state policies and restricted to the private realm of people. State policies were prescribed to be done based on rationality. The concept of sovereignty gave ultimate authority to leaders of states. These developments led policymakers and academic scholars to stay away from considering religion as an important factor. J. Fox and S. Sandler (2004 and J. Fox 2001) explain the reasons for the marginalization of religion from the discipline of International Relations. First, the origin of social sciences was rooted in the rejection of religion. The scientific revolution promoted rational explanation and a positivist framework. Early social scientists rejected religion as an influential factor and even predicted its demise. Second, the theories of International Relations and other social sciences were focused on Western European society, where secularization developed faster than any other part of the world. After European colonization and engagement with rest of the world, social scientists explained the public role of religion in Asia and Africa as primordial and predicted that it would disappear as these societies get modernized. However, ironically modernization caused a resurgence of religion rather than its demise in both the West and the rest of the world. The third reason for the negligence of religion in International Relations was the influence of behaviourism and quantitative methodology in the twentieth century. Religion was considered as a difficult factor to measure in quantitative terms. Fourth, the frameworks of major theories of International Relations were based on assumptions that exclude religions. For example, it is hard to accommodate religion into the realist assumptions of a rational state and international anarchy. However, Nukhet A. Sandal and Patrick James (2010) have explained different ways to incorporate religion into existing theories of International Relations. For Sandal and James, marginalization of religion from the discipline of International Relations was "due to the reluctance of IR theory scholars rather than the nature of IR per se".

Return of Religion

Although religion continued as a powerful factor in many parts of the world, scholars of social sciences and International Relations (IR) had ignored the importance of religion. However, various events across the world made the religious factor more visible and policymakers and scholars realized the role of religion in shaping the behaviour of people and policies of states.

There were various factors behind this return of religion. One of the major reasons was recognition of the non-western world as equal sovereigns in international relations. After decolonization, Asian and African countries became more visible at the international level. Scholars from these countries questioned the Eurocentric Orientalist framework of western scholars. Deviating from conventional Eurocentric and ethnocentric analysis, western social scientists also started to study rest of the world with respecting their cultural differences. In this sense, the "return of religion" is not just because of a change in the relation between religion and politics, but it is also because of a change in the frameworks of analyzing it. Religion had continued as an important factor in the public sphere in many parts of the world. Michael C. Desch (2013: 31) argues that most of the world outside of the Atlantic was not secular in a meaningful sense.

Religions like Hinduism and Islam had not accepted the western version of separation of religion from the public arena. These religions were powerful motivations of freedom struggle in many African and Asian countries. The Orientalist scholars who considered religion as a pre-modern feature of society expected that its importance would decline as society got modernized. However, the experience in the following decades proved that their expectation was wrong. Gradually social sciences recognized that the experience of Western society in its progress from religious to secular is not a universal phenomenon, and modernization in non-western society does not necessarily result in the rejection of religion.

During the Cold War, some changes in global politics attracted the attention of scholars to considering religion as a factor in international politics. For example, the Islamist movement became powerful in West Asia after the defeat of secular leader Jamal Abdul Nasar against Israel in the 1967 war. Michael C. Desch (2013: 27) quotes Daniel Philpott, arguing that the "six days war of 1967 was the significant beginning of a resurgence of religion. It awakened religious consciousness among Israelis and crippled the prestige of secular nationalism among Arab Muslims". With the defeat of Arabs in the war of 1973 and the Camp David Treaty of Egypt with Israel and the US, the leadership of Arab states shifted from Egypt to Saudi Arabia.

Events such as the Islamist revolution in Iran in 1979, support of the US to non-state Islamic militants against the Soviet invasion of Afghanistan, the election of Ronald Reagan as president of the US with the support of politicized Christians, Jews and Mormons, and success of Polish Solidarity Trade Union against the Communist state with the support of Chaotic Church and Pope John Paul II, Sikh movement of Khalistan in India and violent conquest of Golden Temple, the rise of Hindu nationalist Bharatiya Janata Party (BJP), shifting of power from secular movement to Islamist group in Palestine after first Intifada, and rise of the Islamic State(IS) as a threat to existing nation-state order are some examples of the resurgence of religion in western and non-western states.

In a demographic sense, new studies indicate that the population growth rate of believers in the world population is higher than non-believers. Major religions including Christianity (1.27% p.a), Islam (2.11% p.a) and Hinduism (1.54% p.a) have higher population growth rate than that of world total population (1.22%). At the same time growth rate of "nonreligious" (0.8%) and "atheists" is at the slowest rate (S. Desch 2013: 22)

Although religion became more visible in global politics and many branches of social sciences considered it as an important factor, the discipline of International Relations (IR) continued ignoring religion in its analysis. During the Cold War, the discipline was more focused on power politics and balance of power. After the end of the Cold War, some scholars turned to religion and explained the importance of religious identity in future global politics. For example, the "Clash of Civilizations" thesis of S. P. Huntington predicted that religion would be the base of conflict in the coming years. However, J. Fox and S. Sandler (2004:1) argue that despite Huntington's *Clash of Civilizations* and *The New Cold War* of Jurgensmeyer (1993) were published in the 1990s, the discipline of International Relations was not ready for the inclusion of a religious variable into its paradigm. Even the constructivist theory, which can accommodate religion into its framework easier than other theories, ignored the role of religion in creating identity and norms. Timothy Shah noted that "religion has become one of the most influential factors in world affairs in the last generation but remains one of the least examined factors in the professional study and practice of the world" (Quoted in M. S. Desch 2013: 14).

However, by the 9/11 terrorist attack on the US, scholars realized the necessity of understanding religious influence. It exposed the limitations in IR theories in analyzing and forecasting the role of religion in international politics. L. M. Herrington and A. McKay (2015:4-5) paraphrased Daniel Philpott arguing that "9/11 emphasized, possibly more than any other recent historical event, that religion continues to be a potent force in global politics." The mainstream journals published more literature on the role of religion in International Relations, Security and War than in previous decades. M. Herrington and A. McKay (2015:6) argue that more books were published about Islam and war in the decade after 9/11 than from the invention of the press in the fifteenth century to 2001 on the subject (154). Ron E. Hassner (2013: 68) says that the number of books in the Library of Congress catalogue on religion and violence has increased from two or three books per year in the last three decades to fourteen books per year since 2001. However, the focus of the literature published after 9/11 was on the negative aspect of religion and its influence on non-state terrorist groups. Attention given to the role of religion as a promoter of peace in international relations has been negligible.

However, it is to be noted that the 9/11 attack on the US was only an eye opener for scholars to consider the role of religion in international politics seriously. Religion was an influencing force even before this attack since it had influenced policies of both state and non-state actors. It was a source of motivation and norm for peace activists on nuclear and other issues. The following section elucidates the ways in which religion influences international politics.

Influence of Religion in International Relations: Theoretical Explanations

The role of religion in international politics can be explained by various IR theories. Religion influences each of Waltzian's three levels of international relations: individual level, state level, and system level. It shapes the personality and preferences of policymakers; it influences the domestic regime and interests of states, and it is a major source of international norms. J. Fox and S. Sandler (2004) identify many influences of religion in international relations. First, it influences the worldview of individuals and, therefore, shapes their thoughts and behaviour. Second, it is one of the bases of identity. Third, it is a source of legitimacy. Fourth, it is associated with formal institutions that can influence the political process. Even though Karl Marx did not give much attention to religion, his statement on religion as the opiate of the masses refers to the influence of religion on the thoughts and behaviour of people. Religion can shape public opinion on state policies related to war. After analyzing the relationship between religion and public opinion on security, James L. Guth (2013: 179) concludes that religious belief makes a difference in public opinion. People's belief influences how they respond to a state's security policy.

These influences are not necessarily unidirectional. They can lead to both peace and war. They can be a legitimate source for keeping the power of government and also be a motivation for the masses to fight against the government. Religion can be a supporter of or a constraint to policies of governments, and it can be a source of normative argument for both government and opposition parties. Therefore, religion plays a dual role in international politics. Even though rulers of a state do not have a monopoly over the interpretation of religion, they try to convince people of the legitimacy of their decisions by using religion. So, religion can be a real motivation and cause for a government's actions or a tool to legitimize an action that has been taken with hidden interests. For example, religion is used to support and oppose Israel's policies' legitimacy in its occupied territory. However, this dual role of religion does not necessarily mean that it can be discarded just because it is used as an instrument by different people with different intentions. Religion is used as an instrument because of its power to mobilize people and get international and domestic support.

Timothy Samuel Shah (2013) has pointed out various contributions of religion to international politics. First of all, the international system of sovereign states itself was ideologically assisted by Protestant Christianity. Protestantism also played a role in subordinating religious authority to political authority. Religion was also a major inspiration behind the development of nationalism in many states. Faith-based nationalism was a powerful source behind the formation of states in Europe in the sixteenth, seventeenth and eighteenth centuries. Another normative impact of religion is its code of conduct to govern conflict and warfare. The concept of 'just war' is an example. Now, these religious norms are embedded in international laws of war. Religious-motivated groups were active in international movements against Weapons of Mass Destruction, including nuclear weapons. Religious principle of prohibition of indiscriminate killing of non-combatants was used to mobilize popular support for movements against nuclear weapons. The recent development of norms of global humanitarianism was also a faith-inspired development. Most of the religious traditions, like Buddhism, Christianity, Hinduism, Islam, Jainism, and Judaism, have contributed to the development of global humanitarianism. The religion-based principle of nonviolent resistance of M. K Gandhi is also an example of the contribution of religion to international politics.

At the international level, religion is also a powerful tool to persuade others. It is used to get support from other states for peaceful or violent policies. During Iraq invasion, G. W. Bush used words with religious imageries such as 'axis of evils' to get international support. Iran characterizes America as a 'great Satan' to mobilize domestic and international support against the West. The speech by Barak Obama in 2009 at Cairo University also demonstrates how religion can be used for public diplomacy at international level. As M. Herrington and Alasdair McKay (2015: 3) say, whether one likes it or not, "the dance between religions and world politics will continue" in the future as well. Therefore, it is necessary to IR theories to accommodate religion to their framework to analyze new international phenomena. J. Fox and S. Sandler (2004: 163) argue that the existing body of International Relations scholarship will be partially flawed unless it does not take religion into account. Therefore, the following section analyzes the possibility of accommodating religion into frameworks of major IR theories such as Realism, liberalism, and constructivism.

Religion and Realism

Realism is a dominant theory of International Relations. Its origin can be traced to works of Thucydides, Niccolò Machiavelli, and Thomas Hobbes. It was developed in the twentieth century as a reaction against Utopianism that failed to prevent or predict Second World War. The fundamental principles of realism are statism, anarchy, survival, and self-help. The state is the legitimate representative of the people, and it is the supreme authority to use violence. International organizations are tools of great powers for their national interest. So, unlike domestic politics, international politics is anarchy. Therefore, it is the duty of statesmen to ensure their national interests and the survival of their states. Realism proposes state leaders to focus on national interests rather than ideology or morality.

Realists adopt a pessimistic approach to international relations. They are sceptical of the existence of universal moral principles and warn state leaders against adhering to any notion of ethical conduct at the cost of national interest. Realism generally does not give much importance to morality or religion in international politics. However, as explained in this section, religion's role can be analysed using a realist framework.

Classical realism, which begins with Thucydides and developed by Machiavelli and Morgenthau, focuses on human nature. The drive for power and the will to dominate are fundamental aspects of human nature. However, unlike structural realism, classical realism agrees that changes at the individual level will affect international politics. For example, the change in foreign policies of Iran in recent years can be explained by personal differences between Ahmadinejad and Hassan Rouhani. Therefore, the religious views of statesmen can influence their policies and, as a result, national and international politics. This influence may be in terms of identity, prestige or norms. For example, Jacques E. C. Hymans (2006) explains the nuclear policies of states based on the psychological aspects of state leaders. He explains how the National Identity Conceptions (NICs) are differentiated from one leader to another and how do they influence the nuclear policies of states. Religious identity and emotional feelings such as religious pride can influence NICs and policies of leaders. Daniel L. Byman and Kenneth M. Pollack (2001) demonstrate the impacts of changes in the personality of statesmen on international politics. They show that people's identity, ideas, and beliefs shape their behaviour and influence international politics. Religion is one of the major factors in shaping these ideas and individual identity.

The state leaders' drive for power, as classical realism explains human nature, can lead them to use religion in their policies especially if the state is based on a religious society. The religious dimension of policies helps leaders to justify their policies and sustain their power. At the same time, if leaders act against the religious interest of society, it may cause their removal from power. The example of Anwar Sadat, who was murdered after signing the Camp David Agreement with Israel and the US, shows that even dictators cannot act against the religious interest of society in the name of "national interests". Therefore, statesmen may have to make policies according to religious interests and norms of society. However, the policies of states cannot explain purely based on domestic culture. For example, as Nukhet A. Sandal and Patrick James (2010: 10) indicated, the policies of Iran cannot be explained only by looking its nature of an "Islamic state." The individual views of Ahmadinejad and Hassan Rouhani's beliefs and their interpretation of religion are also important in understanding the policies of Iran.

Sandal and James (2010: 10-11) say that these concepts are flexible even though classical realism talks about power and interest. He quotes Morgenthau supporting his argument that interest formation depends upon cultural and historical context. So, power and interests do not necessarily mean military power or military interest. Religion, as a significant element of culture, can influence in defining interest. In a particular society, if religion is a powerful tool for bringing people together, it is also a part of interest. Sometimes political leaders may use religion as a tool, as instrumentalism explains, for achieving their goals. So, the theory, which is based on the human nature of the desire for power, can accommodate religion both as an independent and intervening variable.

Structural realism, which is developed by Kenneth Waltz, proposes that the struggle for power in international politics is not because of the human nature. The nature of the international structure of anarchy and inter-state competition for the relative distribution of power cause international conflicts. In this anarchical global order, states have no option other than self-help by increasing their power. The changes in human nature or domestic regimes will not change this international structure. So, the theories based on the state level or individual level cannot explain international relations. Kenneth Waltz calls them reductionist theories.

Structural Realism does not give much relevance to cultural variables. Therefore, Structural Realism, especially Waltzian framework of realism which considers state as a "black box" and explains its behaviour based on the concept of "balance of power," is the most difficult theory to accommodate religion. At the same time, Stephen Walt's explanation of the "balance of threat" can accommodate religion into its framework. Because religion is a crucial factor in considering other powers as friends or foes. For example, the rivalry between Iran and Saudi Arabia or the conflict between Arabs and Israel cannot be explained without considering religious factors.

Some realist scholars like Fareed Zakaria distinguish between power and influence and argue that states try to increase their influence in other states rather than power (Sandal and James 2010: 14). Religion is an important factor for a state to spread its influence among other states by legitimizing its actions and by using religion as a tool for soft power. Spreading religious ideas to other states increases a state's influence among them, and this becomes a potential pool for alliance formation (Fox and Sandler 2004: 168). Religion facilitates great powers to create international norms to legitimize their activities and to spread influence among others. The diasporas belonging to the same religion is another power resource of states. For example, the Jewish Diaspora in the United States helps Israel to influence foreign policies of the US.

Religion and Liberalism

The roots of liberalism can be traced to enlightenment era and the works of Immanuel Kant. Unlike realism, liberalism does not consider war as an inevitable nature of international relations. War can be overcome through international trade, international organizations, and democratization of states. Even though liberalism emphasizes on the possibility of cooperation, the diverse schools within the theory suggest different ways of achieving it. The focus of some schools, like commercial peace theory, is on economic aspects of international relations. Therefore, these strands of liberalism are less accommodative to religion into their frameworks. There is also criticism against liberalism on its presumption about religion as a source of violence and the modern international system as a promoter of peace and human rights. Philips Gorski and Gulay Turkmen-Dervisoglu (2013: 136-144) call these concepts of liberalism as myths and argue that religion is not inherently violent nor is secular nationalism inherently peaceful.

Compared to other strands of liberalism, neoliberalism is more amenable to religious variables. The basic assumptions of the theory are different from those of realism. First is its pluralist approach. It is not just states, but non-state actors are also important in international politics. Second, state is not a unitary actor. Liberalism does not consider state as a black box. According to liberalism, individuals have their own interests. The preferences of states derive from internal dynamics and bargaining process. State policies are results of competition among different kind of decision-makers. Third, states have multiple agendas. Unlike realism, which gives preference to security agenda of states, liberalism recognizes multiple agendas, including economic, environmental, and other interests of states.

These assumptions of neoliberalism provide more space for religion in international politics than realism. Since it considers non-state actors a significant force in international politics, it can explain the role of religious organizations and networks such as al-Qaeda and the Catholic Church in international politics. Globalization and development of communication and transport systems allow more interaction among people with same religion and help in the development of transnational religious networks. These religious groups are autonomous from state authority and work with their own agendas. Their direct interaction with people influences public opinion and so policies of states. The domestic religious pressure groups play an important role in the decision-making process of states. Since the theoretical assumptions of neoliberalism do not consider states as a "black box" and recognize the bargaining process within states, it can cover the role of these domestic groups. The assumption of multiple agendas allows states to think beyond security issues. Just like economic and environmental agendas, religious agenda can also be a preference of states.

According to neoliberalism, international organizations and regimes have impact on the policies of states. For example, the 'human rights regime' has influenced international relations by supporting humanitarian intervention and by constraining human rights violations. At the same time, religion can strengthen and weaken these regimes. In the case of human rights, religion is one of the sources of human rights, and it also constrains the universalization of human rights, like the rights of women, in the name of particularism and cultural differences. The framework of neoliberalism is useful to understand the role of "just war tradition" in the formation of international humanitarian laws and to analyze its influence on international wars. Religion can also be a base for international organizations like Organization of Islamic Conference (OIC).

The idea of "soft power' is another concept in neoliberalism to understand the influence of religions. Liberalism does not restrict the concept of power to military power. Power of a state is its ability to influence the behaviour of other states and of its own people. This influence can be by using hard power such as the military and economy. It also can be by soft power of attracting other states through its culture, political value and its policies resting on legitimacy and moral authority (Nye 2004: 11). Religion is a powerful element in these resources of soft power. In many countries, religion is a primary source of culture. Huntington considered religion as the main pillar of almost all civilizations. States use religion to spread influence among other states with the same religious background. For example, Buddhism is a powerful element in the soft power of India, especially in its relation with East Asian countries. Saudi Arabia uses its position as the custodian of holy places to spread its influence among other Muslim countries. In recent decades, even non-state militant groups tried to use religion as a soft power for influencing people. Religion is also an important source of political values. For example, Gandhi developed the idea of nonviolence from his religious backgrounds. As it was explained earlier, religion is a powerful tool to legitimize actions of governments and to get moral supremacy both at domestic and international levels. Fox and Sandler (2004: 39) suggest that religion can legitimize what politician cannot achieve by other means. This legitimacy and moral supremacy are useful for convincing others the validity of state's actions.

Religion and Constructivism

The origin of constructivism can be traced into the 1980s as a reaction to mainstream IR theories of neo-realism and neo-liberalism. Constructivism focuses on how ideas, rather than material factors, are important in shaping the identity and interests of states. Constructivism argues that the structures are determined by shared ideas rather than material forces. Alexander Wendt (1992) argued that "anarchy is what states make of it." Identity and interests of states are inter-subjective constructions rather than predetermined ones. The enmity and friendships are determined by interactions, and "security dilemma" of realism and "security community" of liberalism are determined by this identity of enmity and friendship.

Since Constructivism focuses on ideas and identity, it is the most potential theory to accommodate religion. However, despite this theoretical possibility of accommodation, constructivist theorists have not given much attention to the role of religion in the development of the ideas or identity in their practice of the theory. Religion is a powerful force in shaping ideas and beliefs of people and to determining what is right and what is wrong. Religion suggests moral principles and norms for interaction among people and nations. It puts forward many conditions and restrictions with regard to warfare. Religion determines ultimate goals of each including political leaders. Therefore, in a religious sense, a successful political leader is one who follows these religious principles to achieve their ultimate goal of success. Many religions do not support the "dual morality" proposed by realism. The morality of leaders of states is not differentiated from individual morality. The idea of separation of religion and state is also rooted in religious principles. When Protestant Christianity was the promoter of this separation in the seventeenth and eighteenth centuries, Islam did not separate religion from politics. So, the orientation of leaders regarding their duties and goals are influenced by their religious background.

Religion also influences in determining friends and foes. Differences in ideas lead to different identities and interests. The conflict between

Israel and Arabs cannot be explained without considering their religious backgrounds and the role of religions in creating ideas about the territory of state and the importance of holy places like Jerusalem. At the same time, the support of many Muslim countries to Palestinian freedom struggle is also influenced by religious ideas and identities. Scott Thomas (2000: 4) says that "a shared identity produces a sense of psychological affinity while conflicting identities produce a sense of psychological distance." Constructivism explains the actions of states as a result of their interests based on their identity. The difference in identity, such as theocratic or secular and Islamic or Christian, leads to a difference in actions. An analysis of these identities is necessary to understand the actions of states.

Religious principles of universal humanism and brotherhood have impacts on the humanitarian assistance of one state to another. Religion promotes norms of humanitarian assistance irrespective of international borders. However, the selection of beneficiaries of this aid has also been influenced by religious identity. Martha Finnemore (1996: 135) says that "from the nineteenth century, religion seems to be important in both motivating humanitarian action and defining who is human." Even when Christian charity concept promoted this assistance, Christian states were privileged as beneficiaries over Muslim states.

Religious identity works as a base for nationalism which is defined as a feeling of belongingness to an "imagined community" (Benedict Anderson 1983). The Christian nationalism in Poland, Islamic nationalism in Pakistan, Hindu nationalism in India and Jewish nationalism in Israel are examples of nationalisms constructed on the basis of historical and religious imaginations and ideas. However, the idea of nation-state, which means the creation of states based on particular identities, is a modern secularist idea. Although religion can be a base for an identity, it does not propose to create a separate state exclusively for this identity. However, in this era of the nation-state, religion plays an imperative role in developing nationalism.

Even though constructivism focuses mainly on the identity of states, the feeling of a particular identity creates a lot of psychological impacts on the individual as well. The feelings of status and pride of one's religion can motivate individual leaders to act beyond the expectations of material rationality of realism. This psychological feeling was one of the motivations of Z. A. Bhutto in his statement on Pakistan's nuclear weapon as an 'Islamic Bomb'.

Religion is a key force in the constructivist concept of 'logic of appropriateness'. As Michael Barnett (2011: 155) said, "the 'logic of appropriateness' highlights how the actors are rule-following, worrying about whether their actions are legitimate." Political leaders can get legitimacy of their action by using religious ideas and norms. However, this legitimacy depends on whether the international and domestic audiences share the same idea of these actors. Even people belonging to the same religion may have different ideas and interpretations on an issue.

Copenhagen school of constructivism can also accommodate religion in its concept of 'securitization'. Political leaders can securitize a particular religion as a threat to states or culture. For example, S. P. Huntington's "clash of civilizations' thesis has enforced Islamophobia in western countries. Quoting Waever et al. (1993), Lee Marsden and Heather Savigny (2013: 208) say that "religion and national identities should be considered as social security rather than individual one. So, society itself is as being under threat from alien 'other'". So, Islamic ideology and its spread have been securitized in western society. Religion is also a useful tool to securitize one group of people or other states as a threat the nation. The attacks on minority communities can be explained by using the framework of Copenhagen school of constructivism.

Islam and Nuclear Weapons

Since this study focuses on Iran and Pakistan as cases, an understanding of the Islamic view on nuclear weapons is inevitable. Therefore, this section analyzes different views within Islam on the acquisition and use of nuclear weapons for first and second strike purposes. The understanding of the Islamic view is helpful to analyze the nuclear behaviour of fifty-six member states of the Organization of Islamic Conference (OIC) out of which fifty-five have signed Non-Proliferation Treaty (NPT), and forty-nine have signed Comprehensive Test Ban Treaty (CTBT) (Sohail H. Hashmi 2004: 345). Pakistan is the only Muslim majority country which has not signed both NPT and CTBT, and Kazakhstan is the only Muslim majority country which gave up nuclear weapons after possessing it at the time of disintegration of the USSR. However, deep empirical and theoretical knowledge is needed to check the role of Islam in these decisions.

Even though nuclear weapons were not introduced during the time of Prophet Muhammad, who propagated Islam and established an Islamic state in the seventh century, the basic principles to deal with a similar situation were prescribed. Islamic jurisprudence developed ethics and

41

conduct of warfare from these basic principles of the *Quran* and *Hadees* (words and practices of the prophet). Along with the *Quran* and *Hadees, Ijmaa* (unanimous opinion of companions of the Prophet or scholars) and *Qiyas* (analogy, i.e., rational interpretation of the new situation in the light of other sources) are basic sources of Islamic jurisprudence. Along with others, *Qiyas* is the significant source of the laws on nuclear weapons. However, as John Kelsay (2006: 81-85) correctly pointed out, Shari'a (Islamic Law) reasoning and interpretation are not easy tasks. It requires a mastery of Arabic language, especially special terms in the Quran and Hadees, knowledge in the context of each verse of the Quran and Hadees, and clarity in particular and general meaning of the verses. Due to these difficulties, Shari'a reasoning is restricted to the learned people of "*Ulama* (Scholars)".

The ignorance of these criteria and restrictions leads scholars like Rolf Mowatt Larssen (2011) to consider the fatwas of Bin Laden and Al-Zawahiri as major opinions in Islamic jurisprudence on nuclear weapons. Sohail Hashmi (Sohail H. Hashmi 2004: 322) also made the same mistake when he classified the contemporary Islamic view on the ethics of WMD into three and counted Muslim WMD terrorism as one of them. As James Turner Johnson (2011: 60) correctly said, "Bin Laden and his associates who signed the *Declaration* and issued it as a fatwa lacked mandatory religious authority to issue a such statement of religious judgement". John Kelsay (2006: 104) also elucidated that signatories of *Declaration* or author of *The Neglected Duty* are unqualified. Because only learned *Ulama,* who have completed an advanced course of training, have the right to issue an independent opinion.

Shari'a laws deal with all activities of individuals and classify them into five broader categories: *Vaajib, Musthahabb, Mubah, Makruh, and Haraam.* As explained by figure-2, *Vaajib* means a necessary or obligatory action. For example, *Zakat* is an obligatory action for all Muslims who meet the necessary criteria of wealth. *Mustahabb* indicates a recommended action in Islam, but without compulsion. Islam motivates to do *musthahabb* and offers rewards for it, but one person can also avoid it. *Musthahabb* prayers which is promoted in addition to five-time compulsory prayers are examples of it. *Mubah* refers to an action that is neither commanded nor prohibited by Shari'a like drinking water. *Makruh* means an action that is discouraged by Islam, but has not been prohibited strictly. Using an extra amount of water for ablution (*Vudu*) is an example of *Makruh. Haraam* refers to forbidden activities such as drinking alcohol. While Vaajib and *musthahabb* mean action is encouraged, Haraam and Makruh mean an action is discouraged. The possession and use of nuclear weapons can also be judged by using this classification. Various views of Islamic scholars on nuclear weapons can be analyzed by using these five categories. This classification is useful to identify whether Islam, according to each opinion, *de jure* motivates nuclear weapons or not.

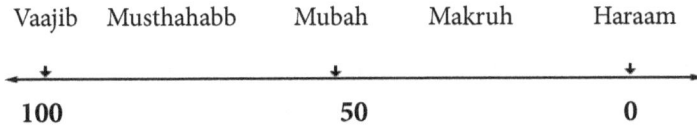

Vaajib	Musthahabb	Mubah	Makruh	Haraam

100	50	0

Figure-2

Basic Classification of activities in Islamic Jurisprudence and degree of motivations of motivation

Sohail Hashmi (2004: 322) classifies the Islamic view on the ethics of WMD into three categories. First, 'WMD Jihadists', who support the acquisition and use of WMD in the right circumstance if conditions are met. Second, 'Muslim WMD terrorists' who argue that the acquisition of nuclear weapons is necessary and support their use as a first resort. Third, 'Muslim WMD pacifists' those who oppose acquisition and use of WMD, considering them to be against Islamic ethics. Hashmi supports Muslim WMD pacifists because of many reasons. First, these weapons do not discriminate between combatants and non-combatants which is compulsory under Islamic laws of war. Second, even if WMD is used against the military exclusively, that will kill them in a horrible way that is against Islamic teachings. Third, WMD creates environmental damage, and that is discouraged by Islam even in wartime. Fourth, since the WMD cannot be used for any morally justified purpose, the resources diverted to them are equal to what Quran called as *israf* (waste). One more reason for opposing the acquisition of WMD for the deterrent purpose is, if the enemy of a state is sure that use of nuclear weapons is prohibited by religion and that the state would not use it, then deterrence would be flawed and would not be credible.

Out of these three views on nuclear weapons, WMD terrorists consider acquiring the weapon as necessary. While WMD jihadists consider acquisition as *Mubah* and its use is *Mubah* if only conditions are met. Use of WMD as first strike weapon is not allowed and can be labelled as *Haraam*. WMD pacifists consider both acquisition and use of nuclear weapon as prohibited by Islam. Out of these three views, religion can motivate nuclear weaponization only according to nuclear terrorist's interpretation. However, as explained earlier, terrorist leaders such as Bin Laden and Al-Zawahiri are not qualified for giving an independent opinion about Islam. According to the view of WMD jihadists, the religion allows nuclear weaponization. However, a *Mubah* motivates neither an action nor inaction. As per the interpretation of WMD pacifists, the religion motivates non-weaponization and promotes non-proliferation.

Rolf Mowatt Larssen (2011) figures out various opinions within Islam on nuclear weapons. Mistakenly, he also considers the opinion of terrorists as equal to the opinion of qualified scholars. According to Larssen, terrorists motivate acquisition and use of nuclear weapons. However, as Larssen himself explained, both Sunni and Shia scholars who have influence in the Muslim societies and state policies have either prohibited nuclear weapons or allowed with certain conditions. Since they can be leveled either as *Mubah* or as *Haraam*, according to neither of these two opinions does religion motivate nuclear weaponization. It should also be noted that none of the Shia leaders with the rank of Majra al-Taqlid has allowed nuclear weaponization. In Shia Islam, *Marjaal-Taqlid,* literally meaning "source to imitate/follow" or "religious reference", is a title given to the highest level Shia authority, a *Grand Ayatollah* with authority to make legal decisions within the confines of Islamic law for followers and less-credentialed clerics.

At the same time, according to those who prohibit it, religion demotivates nuclear weapons. Larssen (2011: 44) says:

> "Fortunately, this consciousness introduces an additional level of scrutiny over the wisdom of the use of WMD in the Islamic world that does not exist in secular states that are under no obligation to seek any form of religious or moral authorization for their use."

Among Sunni scholars, Ali Gomaa, the Grand Mufti of Egypt, has given a comprehensive fatwa against nuclear weapons. Briefly, he opposes WMD including nuclear weapons for the following reasons: First, the use of WMD is a violation of international treaties. Second, it kills people by surprise without warning. Third, Islam forbids harming of women and children. Fourth, the killing of the Muslim population in targeted countries is prohibited. Fifth, use of WMD will cause a catastrophe to the entire world. Sixth, it damages individual and public properties. Seventh, the *Qiyas* (analogy) of using WMD to *tabyit* (night attack) or using a catapult.

Muhammad Tahri-ul-Qadiri (2010), a Pakistani scholar, also has given one of the most comprehensive fatwa against terrorism which includes fatwa on the ethics of warfare. He also bans the indiscriminate killing of non-combatants and harming of women and children.

Among Shia scholars, Ayatollah Ali Khamenei, the Iranian supreme leader, is "said to have issued a fatwa against nuclear weapons" (Larssen 2011: 50). The other Shia scholars who prohibited nuclear weapons include Ayatollah Ruholla Khomeini, who is the founder of the Islamic revolution and the highest ranking Maraja in recent decades, Grand Ayatollah Yusef Saanei and Grand Ayatollah Hussein Ali Montazeri. The Shia scholars who believe that nuclear weapons could be considered under a certain condition, such as Ayatollah Muhammad Taghi Mesbah Yazdi, Mohsen Gharavian and Mohsen Kadivar (Larssen 2011: 53) also do not consider nuclear weapons as the duty of state leaders. However, since they are not in the position of Majra al-Taqlid, they are not authorized to issue a fatwa on this matter, and they do not have followers. In short, neither recognized Sunni nor Shia scholars consider nuclear weaponization as the duty of state leaders.

Chapter III

Role of Religion in Nuclear Policies of Iran

Introduction

Iran is a country with technological capability for the nuclear fuel cycle. This technological development makes the international community concerned about the potential nuclear weaponization of Iran and its proliferation. Scholars belonging to different theoretical backgrounds have analyzed the motivations of Iran for nuclear weaponization. Some scholars start with the presumption that Iran has already decided to make a nuclear bomb and then try to explain why does Iran proliferate? For example, Michael L. Farmer (2005) titled his work "Why Iran Proliferate?". Saira Khan (2010) analyzes multiple sources of security threats and protracted conflict as the motivation behind Iran's weaponization. Michael Clarke (2013) forwards concepts of "status" and "security" along with isolation from existing international order (he uses the term 'pariah state') as the reasons behind acquiring nuclear weapons.

However, one fact remains, as Christopher J. Bolan (2013) said and as US National Intelligence Estimate of 2007 and 2011, and all reports of the International Atomic Energy Agency (IAEA) confirmed, that Iran has not yet developed its nuclear weapons, and during 13 years (2003-2017) of the most intrusive inspections ever made in the history of the IAEA, the Agency did not find any evidence of diversion of the Iranian nuclear program toward weaponization. Officially, Iran argues that Weapons of Mass Destruction, including nuclear weapons, are against the principles of Islam, which is Iran's state religion and prohibited by the fatwa of the Supreme Leader Ayatollah Khamenei. Religious prohibition is presented as a constraining factor in nuclear weaponization. This case study on Iran investigates the role of religious principles in the nuclear decision-making of Iran. The first part of the chapter summarizes the historical developments

46

of the nuclear program in Iran under the Shah and the Islamic Republic. The following section analyzes the motivating and de-motivating factors for the nuclear weaponization of Iran. This section reviews explanations of various theories regarding the nuclear weapon program of Iran. The next section examines the role of religion in its nuclear policies. The section goes through the position of the Supreme Leader and other religious scholars on a nuclear weapon and analyzes the importance of their position in the nuclear policies of the state.

History of Nuclear Iran

Nuclear program in Iran started in the 1950s with a nuclear cooperation agreement signed with the US in 1957 under Atom for Peace program. Following this Agreement for Cooperation Concerning Civil Uses of Atoms, Iran acquired Tehran Nuclear Research Centre with a research reactor and highly enriched uranium to fuel the reactor (Maurer 2014: 50). Mohammad Reza Shah Pahlavi, the Shah of Iran from 1941 to 1979, was very much interested in modernizing Iran through technological development including nuclear technology. The oil price boom in 1973 following Arab-Israeli War increased the oil revenue of Iran from less than $1 billion to $20 billion by the end of the 1970s. It enabled the Shah to develop the scope of his nuclear plans. In March 1974, he announced an ambitious nuclear plan for developing 23,000 MW (e) of nuclear power capacity and for constructing over 20 nuclear power reactors. For institutionalizing his plan, Shah established the Iran Atomic Energy Organization in the same year.

Shah benefited from his relation with western countries by getting their support for the program. In fact, the US laid the foundation of a nuclear Iran beginning in 1957 (Rowberry 2013). The US built the first nuclear facility in Iran, Tehran Research Reactor (TRR) in 1967 and provided Iran with fuel for that reactor — weapons-grade enriched uranium (Rueters 2012). In 1976, President Gerald Ford issued a directive for Iran to have full fuel cycle including a US-built reprocessing facility for extracting plutonium from nuclear reactor fuel (Linzer 2005). The Ford strategy paper said the "introduction of nuclear power will both provide for the growing needs of Iran's economy and free remaining oil reserves for export or conversion to petrochemicals" (Linzer, D. (2005).

A 1974 CIA proliferation assessment stated, "if [the Shah] is alive in the mid-1980s... and if other countries [particularly India] have proceeded with weapons development, we have no doubt Iran will follow suit" (SNIE

4-1-74 1974: 38). Despite this fact, Washington decided to support a nuclear Iran. The USA agreed in 1975 to construct eight nuclear reactors. The Massachusetts Institute of Technology (MIT) signed a contract for training Iranian nuclear engineers.

The Shah had also received the help of other Western countries. For example, in 1975, Shah signed a deal with Kraftwerk Union AG, a German company, to build its first power reactor at Bushehr. French company Framatome agreed to build two reactors (Kibaroglu 2006: 214-215). Saira Khan says that Iran obtained 22 reactors for generating 23,000 MW of electric power from the agreements with Germany, France, and the US. In 1974, a joint Iranian-French company was established called "Sofidif" where France and Iran owned 60 and 40 percent, respectively. The purpose was to supply nuclear reactors, enriched uranium and a nuclear research centre for Iran. Later, Sofidif purchased 25 percent of the Eurodif share, a consortium operating a uranium enrichment plant in France, giving Iran its 10 percent share of Eurodif. Iran paid $1 billion in 1975 and $180 million in 1977 in return for the right to 10 percent of the company's LEU production (Sahimi 2010).

According to Saira Khan "the Western allies had helped Iran in developing a comprehensive nuclear program under the umbrella of Cold War alliance with the understanding that Iran will never have the ambition to acquire nuclear weapons" (Khan 2010: 48). In addition to these countries, Iran benefited from UK and India from where Iranian scientists received nuclear training and South Africa and Algeria from where uranium was imported (Maurer 2014: 50). David Patrikarakos pointed out that "the Tehran Nuclear Research Centre was 'equipped by the United Kingdom' and staffed by British, Turkish, Pakistani and Iranian scientists" (Patrikarakos 2012: 34-35). By the end of Shah's regime, Iran had developed two reactors at Bushehr and had acquired the basis of the civil nuclear program.

The declared goal behind the Shah's nuclear program was to develop nuclear energy for electricity. Nuclear energy was considered as a cheap source of power and a means to improve the country's economic position and standard of living. In 1974, Shah's office announced three economic reasons for nuclear power - "benefit of resource diversification, energy competition, and technological advancement" (Patrikarakos 2012: 25). Shah calculated that oil which is the primary source of national income should not be burned for domestic needs and development of nuclear energy as an alternative source of power would help the country to increase national income by exporting more oil to international markets.

Even though the official position stresses economic motivations, there were also other motivations like prestige for his personality and the nation. Nuclear power was an important source of prestige and a significant factor in dividing the world into modern and modernizing states. Developing countries like India considered nuclear power as a tool to improve status at international level and to free from the legacy of colonization. These countries regarded nuclear power as an important tool to defend against the dominance of developed countries and to avoid dependence upon them by attaining self-sufficiency. However, unlike leaders of India and many other developing countries, Shah had not judged the international politics as a competition between developed and developing countries. For Shah, nuclear power was not a tool to compete with western countries but was a part of westernization of his country and his regional superiority strategy as the US gendarme (Behestani and Shahidani 2015). Shah also judged nuclear technology as a symbol of modernization and as a means to westernization (Patrikarakos 2012: 28-35). He expected that modernization and westernization would bring prestige to his nation.

However, though Shah kept the door open for weaponization, he had no plan to develop nuclear weapons soon. After quoting Shah's statement of 1975, "if other countries in the region acquired nuclear weapons at some point, Iran would be compelled to follow suit", Saira Khan deducted two points: the Shah had no intention for nuclear weaponization, and he was aware of easy option of nuclear weapon if Iran achieved technological advancement (Khan 2010: 12). So, he kept the option for nuclear weapons open and tried to develop full nuclear fuel cycle.

Akbar Etemad, who is known as the father of Iran's nuclear program and was the president of the Atomic Energy Organization of Iran (AEOI) between 1974 and 1978 said in an interview:

> "I always suspected that part of the shah's plan was to build bombs. So, I came up with a plan to dissuade him. I asked the shah if I could spend a few hours every week teaching him about nuclear technology. I thought he should know enough about nuclear energy to know the dangers of a bomb. At the end of the sixth month, I asked him, "So now that you have a good grasp of the technology, what direction do you want to take? Do you want to use it for peaceful purposes or to build bombs? I have to know that in order to plan it. We talked for about three hours, and the Shah told me his ideas about Iranian defence strategy. He thought that Iran's conventional army was already the most powerful in the region and believed that Iran

didn't need nuclear weapons at that moment. He also realized that if Iran developed nuclear weapons, the Europeans and the Americans wouldn't cooperate with it. But I think that if the Shah had remained in power, he would have developed nuclear weapons because now Pakistan, India and Israel all have them" (Bahari, M. 2008).

The Shah was aware of the consequences of developing nuclear arms and its impact on Iran's relation with the US. He calculated that the development of nuclear weapon could isolate Iran from Western powers and it would affect the availability of nuclear technology and uranium from international markets.

Shah had also recognized the shift in global norms on nuclear weapons from a base to be a prestigious superpower and as a sign of status to a base for being a defiant state. The Shah understood that the NPT created a norm of characterizing new weaponized states as 'bad' states. The nuclear weapon was no longer a symbol of prestige (Patrikarakos 2012: 54). The Shah considered nuclear weapons as a less credible deterrent than conventional weapons (Khan 2010: 49). So, he spent the increased oil revenue after 1973 oil price boom for acquiring conventional weapons instead of nuclear weapons. However, based on his talk with Akbar Etemad, Hossein Mousavian, one of the authors of the present book, is convinced that Shah's plan was a full fuel cycle to be prepared to go for nuclear if others like India would go.

Role of Religion in Nuclear Policies of the Shah

The policies of the Shah on nuclear weapons or their use for the civilian purpose were not shaped by religion. At the normative level, he was not against nuclear weapons. The reason behind his refrain from expediting weaponization was his concern about its impacts on the status of his country at the global level and its relation with western countries. Even when he talked about nuclear weapons, he stressed on Iranian or Persian identity than an Islamic one. For example, Patrikarakos points the prestigious feeling of the Shah in his Iranian identity by analyzing Shah's words indicating that if little states in the region acquired the bomb, Iran would reconsider its decision on nuclear weapons (Patrikarakos 2012: 68). The nuclear program for the civilian purpose was also to improve the status of Iranian identity and to bring back the glory of Persian Empire rather than Muslim identity or Islamic Empire. So, religion was not an important factor in the nuclear policies of Iran during the period of the Shah.

Nuclear Iran: Under Islamic Republic

The Islamic revolution of 1979 and the events in the following years including hostage crisis and war with Iraq created dramatic change in foreign and security policies of Iran. The revolution also impacted Iran's nuclear program. For the Shah, the nuclear program was part of modernization and westernization of the country. Modernization was treated as equal to westernization. The first supreme leader of Islamic Republic, Ayatollah Khomeini opposed all kind of what he called *Gharbzadegi* (means west-struckness or *westoxification*) (Patrikarakos 2012: 93). The ambitious Shah's nuclear program was also considered as part of Western influence in the country and as a continuation of colonialism by other means. The official position on nuclear program was declared in June 1980 as:

> "The construction of these reactors, started by the former regime on the basis of colonialist and imposed treaties, was harmful for the country from economic, political and technical points of view, and was a cause of greater dependence on imperialist countries. These countries were stopped after victories of the revolution" (Patrikarakos 2012: 98).

So, the main part of the nuclear program was halted in the first years of the Islamic Republic. After the 1979 Iranian Revolution, Iran decided to cancel or shrink both the ambitious nuclear and military projects of the shah. Iran decided only to maintain the operative TRR producing medical isotopes for medical use and to complete Bushehr power plant which was already paid for, and 90% was completed. Iran had no plan to have uranium-enrichment or heavy water activities inside the country. Tehran had an agreement with the French-based consortium Eurodif, to enrich uranium in France and supply fuel to the Tehran Research Reactor and the Bushehr power plant, bypassing the need to have the facilities in Iran.

Following the revolution and under pressure from the United States, however, the French pulled out of the deal, Germany ceased cooperation on the completion of Bushehr and the US stopped providing fuel for TRR. This event forced Iran to proceed with efforts to reach self-sufficiency to complete billions of dollars' worth of unfinished projects and to ensure that it would have adequate supplies of reactor fuel. West withdrew from all nuclear agreements and contracts and isolated Iran through sanctions and other means (Mousavian 2012). Oliver Schmidt (2008) identified many factors in declining of the program. First, leading scientists left the country in reaction to the revolution. Second, Supreme Leader Ayatollah

Khomeini was not interested in nuclear energy. Third, many western companies stopped cooperation with Iran mainly because of the pressure and embargo from the side of the US (Schmidt 2008: 26).

However, in addition to what forced Iran for sufficiency on fuel production, the war with Iraq (1980-1988) and the silence of the international community and institutions over the chemical attack of Iraq changed calculations of Iran towards its nuclear program. The Iranian leaders understood the importance of modern technologies, deterrence, and self-sufficiency to survive in international politics. Iran lost faith in international moral teachings and regimes. Rafsanjani, who was the then speaker and later president of Iran, stated that:

> "the moral teachings of the world are not very effective when war reaches a serious stage and the world does not respect its own resolutions and closes its eyes to the violations and all the aggressions which are committed in the battlefield" (Khan 2010: 12).

Even though Saddam Hussein initiated the war and violated international norms of chemical weapons by using them in the war even in cities, he obtained international support throughout the war. While Saudi Arabia and other GCC (Gulf Cooperation Council) states provided Iraq with financial assistance, it got weapons from Soviet Union, China, France, Germany, and the UK. Western countries helped Iraq by supplying chemical and biological weapons (Khan 2010: 53). The US provided material and technology for Saddam to use chemical weapons against Iran. "The US knew Saddam Hussein was launching some of the worst chemical attacks in history - and still gave him a hand" (Harris and Aid 2013). Jawad Zarif wrote in the Washington Post:

> "In 1980, in the aftermath of the Islamic Revolution, Iraq's Saddam Hussein launched a war against Iran fully supported financially and militarily by almost all of our Arab neighbours and by the West. Unable to secure a quick victory, Hussein used chemical weapons against our soldiers and civilians. The West not only did nothing to prevent this, but it also armed Hussein with sophisticated weapons, while actively preventing Iran from getting access to the most rudimentary defensive necessities. And during the eight long years that this war continued, the U.N. Security Council did not issue a single condemnation of the aggression, the deliberate targeting of civilians or the use of chemical weapons.

This may have been forgotten by most in the West, but it is not forgotten by our people. They remember the missiles raining down, the horrific images of men, women and children murdered with chemical weapons and, above all, the lack of a modern means of defence" (Zarif, M. J. 2016).

Consequently, Iran learned the following lessons from the war: a war with Iraq is possible at any point of time; the international community or institutions cannot be trusted to get help at the emergency situations; international norms on the use of chemical weapons are very weak; and Iran has to be technologically advanced and self-sufficient. It enforced Ayatollah Khomeini's worldview that international organizations are tools of western hegemony. In short, the important lesson from the war was, as Patrikarakos put it, "Iran could trust only Iran" (Patrikarakos 2012: 113).

The new awareness of the necessity of nuclear technology for self-sufficiency changed the attitude of the Islamic Republic towards the nuclear program. The electricity shortage with rapid growth in the population increased the need for nuclear energy. Iran negotiated with Europeans from the mid-80s to mid-90s encouraging them to complete their contractual commitments on Bushehr and fuel supply. The failure of these negotiations, the changes in domestic politics, with the death of Ayatollah Khomeini and the selection of Ayatollah Sayyid Ali Khamenei, who was president of Iran during the war-time from 1981 to 1988, as new Supreme Leader and Ali Akbar Hashemi Rafsanjani as the new president, renewed Iran's nuclear program. However, the nuclear program under the Islamic Republic was different from that of the Shah in its motivations. In contrast to the Shah's goal of westernization of the country through nuclear technology, the Islamic Republic considered nuclear program as a counter to the West.

The experience of isolation after hostage crisis and during the war with Iraq led the new regime to try to develop a nuclear program indigenously. The new government kept a distance from both East and West blocs of the Cold War and joined with non-aligned countries. The fundamental principle of Iran's foreign policy was "Na Sharq, Na Gharb, FaqatJumhuri-ye Islami": "Neither East nor West, only the Islamic Republic [of Iran]" (Kibaroglu 2006: 215). The Islamic Republic promoted 'nuclear nationalism' among scientists and urged all scientists who left the country after the revolution to return to Iran and serve their nation.

However, it was difficult for the government to develop a nuclear program without any foreign help. In early 1991, Amrollahi, the Vice President of Iran and chief of the Atomic Energy Organization of Iran (AEOI), announced that except the US, Israel, and the 'racist South African regime' the AEOI would undertake nuclear cooperation with any other country" (Patrikarakos 2012: 135). However, western countries were not ready to help Iran especially due to the pressure from the US. Finally, China provided needed technologies in the early years of the 1990s. China had trained Iranian nuclear technicians since 1985. However, as a result of the US-China agreement, the involvement of China in Iranian nuclear program declined in 1997. Then, Russian companies supported Iran and Moscow also offered thermal light water research reactors and natural uranium, and training to 15 Iranian scientists per year (Patrikarakos 2012: 140). Iran sought support from Pakistan and Pakistani nuclear scientist A. Q. Khan from 1980s itself (Bernstein 2014: 77). After the US denied the legitimate rights of Iran for even one nuclear power plant and blocking international market from providing fuel for the 1967 American made reactor in Tehran, the nuclear program for Iranians became a part of their integrity, sovereignty, independence, and confronting the US' bullying strategies.

Now, as far as the nuclear fuel cycle is concerned, Iran maintains mining, milling, enrichment and fuel fabrication capabilities. Whether it is for military or civilian purpose, Iranian leadership wants to be indigenously self-sufficient in all process of nuclear fuel cycle. In 2006, Ahmadinejad announced that Iran "has joined the club of nuclear countries" by successfully enriching uranium for the first time (Khan 2010: 14). His understanding of "nuclear club" was "nuclear technology" while the world's understanding was "nuclear bomb". The sanctions from Western nations and political isolation have slowed down the progress of nuclear program of the Islamic Republic compared to that of the Shah. However, this isolation strengthened the indigenous capability of Iran. Nuclear power now symbolizes the capability of Iran and has become a source of its pride.

Iran and Nuclear Weapons

Officially, both systems of the Shah and the Islamic Republic have argued that the nuclear program of Iran is only for peaceful civilian purposes. However, the reason behind this rejection is different for each regime. The Shah's calculation was mainly in materialistic terms and concluded that nuclear weapons would not increase the security of Iran and it would negatively affect its relationship with the US. The Islamic Republic stresses more on ideological aspects and argues that Iran will not develop nuclear weapons since they are prohibited by Islam, which is the official religion of Iran. Nevertheless, analysts, mainly western policymakers and academicians, are suspicious about the argument of Iran and about the influence of religion in nuclear decision-making.

It is clear that technologically Iran is now capable of developing nuclear weapons if it takes a political decision to do so (Clarke 2013: 494). Michael L. Farmer (2005: 42) argues that

> "Iran has uranium ore mining facilities, uranium conversion facilities, uranium centrifuge facilities, and an indigenous missile industry. This means that Iran can mine, process, convert and enrich uranium sufficiently to use in a uranium-based nuclear weapon".

Oliver Schmidt (2008: 46) also says that Iran possesses the technical means to develop nuclear weapons and to deliver a nuclear warhead. According to supply-oriented theories, Iran would have developed nuclear weapons.

However, it is also clear that until now, Iran has not developed nuclear weapons even though, theoretically, it has enough reasons to develop one. According to the US National Intelligence Estimate of 2007 and 2011, Iran suspended its nuclear weapon program in 2003, and Iranian leaders have not made any political decision to build nuclear weapons. Quoting this report, Christopher J. Bolan (2013: 80-81) calls the accusation of nuclear weaponization as a myth. The International Atomic Energy Agency (IAEA) has repeatedly confirmed the absence nuclear weaponization of Iran (Mousavian and Afrasiabi 2012: 2). No other member state of NPT, has been inspected such as Iran. Bolan says "after literally thousands of hours of international inspections, there is absolutely no evidence that Iran is diverting enriched uranium for a weapons program" (Bolan: 83). So, this section analyzes how different theories explain and predict the nuclear behaviour of Iran. The following part figures out the limitations of these theories to explain the absence of nuclearization in Iran and the importance of 'religion' as a factor to understand the nuclear behaviour the state.

Realism or Security Oriented Analysis

According to security-oriented theories like realism, "states build a nuclear weapon to increase national security against foreign threats, especially nuclear threats" (Sagan, 1996: 55). Considering security environment and threats faced by Iran realist scholars predict its nuclear weaponization. Iran has been faced with substantial security threats since revolution in

1979 such as the Iran-Iraq War, separatist rebellions in Iran's Kurdistan and Khuzestan provinces.

Hossein Mousavian (2017) contends that

> Iran has repeatedly experienced the US and Israeli covert cyber attacks, assassinations and the propping up of terrorist organizations like the notorious Mujahedeen-e-Khalq....... While the Middle East is engulfed with civil wars, terrorism, sectarian conflicts, Iran faces serious threats on its borders, whether from terrorist organizations or Afghanistan, Iraq and Syria or a fragile nuclear-weapons state like Pakistan. As one Iranian military official has said, roughly 60% of Iran's borders are not controlled by the neighbouring country (Mousavian 2017).

Iran has been surrounded by many nuclear weapon states such as Israel, Pakistan, India, the US and Russia. Until the invasion of the US in Iraq in 2003, it has faced a threat from Saddam Husain's Iraq which had been reported as possessing Weapons of Mass Destruction. Iran is also a victim of chemical attacks of Iraq in the war soon after the Islamic Revolution. It also has a territorial conflict with the UAE over small islands inside of Persian Gulf. The ideological difference with Saudi Arabia-led Sunni-Arab countries increases the vulnerability of Iran.

Out of these threats, the US is placed at the top (Khan 2010: 64). The antagonistic relation of the Iran with the US after the Islamic revolution and the hostage crisis continues even after three decades. Toward the end of the war, the United States attacked Iranian oil platforms, shot down an Iranian civilian airliner killing all 299 passengers. The presence of the US military in the Persian Gulf region after intervention in Iraq in 1990-1991 increased concern to Iran. This concern again rose after the US's intervention in Iraq without the approval of the United Nations. It showed power of the US in this unipolar world. George W. Bush counted Iran as part of 'Axis of Evil'. The US was able to overthrow Saddam Husain's regime within 21 days which Iran could not do even after eight years Wars of 1981-1988. The US has continually perused regime change policy on Iran. "Regime change will be necessary before the US and Iran can have substantially positive relations," said the US Secretary of Defense, James Mattis (Read, R. 2017).

Iran is not only surrounded by US military bases, but US-made weapons regularly flow into the region — especially the Persian Gulf. Under Barack Obama, the United States sold Saudi Arabia weapons worth roughly $115 billion , which was more than any previous administration. Donald Trump is poised to outdo his predecessor. Moreover, the unparalleled sanctions Iran has been subject to since 1979 increased even after the historic July 2015 nuclear deal. Moreover, while Trump has decertified the Joint Comprehensive Plan of Action (JCPOA), the debates over the deal in Washington are about whether Trump will decide to leave the deal or not. However, the bottom line for Trump and his allies would be to curb Iran's regional power, influence and leverage, and to stop Iran's economic progress through creating tensions in Iran's foreign relations and blemishing the JCPOA.

For Iran, in a realistic perspective, nuclear weapons are necessary to deter American attempt to regime change. It is also noticeable that North Korea, which occupies the third slot of 'Axis of Evil' has not been attacked. So, in realistic logic, Iran has to follow the path of North Korea to deter the US and to avoid the experience of Iraq and Libya. Libya, which dismantled its nuclear facility and program, trusting the US and the UK, paved the way for NATO to attack and destroy its regime.

Israel, which possesses nuclear weapons and has not signed Nuclear Non-Proliferation Treaty (NPT), is another threat to Iran. After quoting Ray Takeyh saying that "the Islamic Republic perceives a nuclear-armed Israel an existential threat, not just to itself but the entire Islamic world", Saira Khan argues that Israel is not just a military threat, but also an ideological threat to Iran (Khan 2010: 61). Iraq was a major threat to Iran until removal of Saddam Husain in 2003. Saudi Arabia also increases its conventional weapons through multi-billions of arms trades with the US. The fact that these regional enemies of Iran get financial and military support from the US increases concern for Iran. Considering all these security threats, it is not surprising that scholars predict the nuclear weaponization of Iran.

In an article in Foreign Affairs, Kenneth Waltz (2012) asserted that Iran "should get the bomb", making a strong case for Iranian nuclear proliferation. He argues that an Iran nuclear state will bring nuclear balance in the turbulent region and more stability to the Middle East by ending Israel's nuclear monopoly. Waltz contends that "it is far more likely that if Iran desires nuclear weapons, it is for the purpose of providing for its own security" (Waltz 2012).

Although Waltz's arguments make sense, but his understanding of Iran is wrong. On Waltz's logic of balancing, during the Iran-Iraq war, when Saddam Hussein repeatedly used chemical weapons against Iran, it should have developed Weapons of Mass Destruction at least to deter Iraq.

However, under the religious instruction of the late Ayatollah Khomeini, Iran never retaliated against Iraq's chemical weapons attacks on Iranian troops and civilians, which killed 20,000 Iranians and severely injured 100,000 more (Porter, G. (2014). Yet another important point is that Iran's Supreme Leader, Ayatollah Ali Khamenei, the successor to Ayatollah Ruhollah Khomeini, prior to Iran's nuclear crisis, re-confirmed a religious fatwa that explicitly and unequivocally bans the manufacturing, stockpiling and use of nuclear weapons, which are deemed "inhuman" and "weapons of the past."

Just like Kenneth Waltz other scholars like Saira Khan (2010), Michael Clarke (2013), Peter Jones (2012), Charles C. Mayer (2004), Oliver Schmidt (2008), and Michael L. Farmer (2005) have predicted nuclear weaponization of Iran considering its security motivations. However, the question that remains to be answered is why has Iran not yet developed nuclear weapons?

Neoliberal Institutionalism and International Regimes

The international regime is another system level variable in explaining the nuclear behaviour of Iran. The neoliberal institutionalism focuses on the role of NPT regime in determining the nuclear program of Iran. This theory expects "Iran to continue to comply with NPT so long as there are benefits from holding into the treaty's commitment" (Tagma and Uzun 2012: 243). Iran is also party to almost all agreements which restrict the use of poison, chemical, and biological weapons including Hague conventions of 1899 and 1907, Hague Declaration of 1899, Geneva Protocol for the Prohibition of the Use in War of Asphyxiating, Poisonous or other Gases, and of Bacteriological Methods of Warfare of 1925, Non-Proliferation Treaty of 1968, Biological and Toxin Weapons Convention of 1972, and the Chemical Weapons Convention of 1993.

Iran maintains that the world powers have violated all three main objectives of NPT (Mousavian 2012a). The first objective was the complete disarmament of nuclear weapons by the NPT nuclear-weapon States: China, Russia, United Kingdom, France, and the United States, of these, none has fulfilled this commitment, and after more than 40 years, they all possess thousands of nuclear weapons. The second objective was to prevent the spread of nuclear weapons and technologies related to nuclear weapons, but the five permanent members of UNSC continue modernizing their arsenals, delivery systems, and related infrastructure. Meanwhile, India, Pakistan, Israel, and North Korea proliferated after NPT and made nuclear bombs. Except in the case of North Korea, the world powers have established strategic relations demonstrating a welcoming attitude towards proliferators. The third objective of NPT was to enhance cooperation in the field of peaceful nuclear technologies. Based on Article IV of the NPT, all member states party to the Treaty have the right to benefit from the peaceful uses of the atom and urges the parties to cooperate with one another in the fullest possible exchange of nuclear equipment, materials, and information for peaceful purposes. Based on this article, research, development, and use of nuclear energy for non-weapons purposes are the 'inalienable right' of non-nuclear-weapon states. The US and the world powers stopped cooperation on peaceful nuclear technologies right after the Iranian revolution while Iran remained as a member of NPT and the US still continues despite the fact that due to the proper implementation of the JCPOA, the technical disputes between Iran and the IAEA have been resolved.

However, even though the opposition of great powers and IAEA has slowed down the nuclear development of Iran, if it really wants nuclear weapons Iran would develop it, especially considering the technological capability of the country. Furthermore, a continuation of Iran in these treaties is due to its commitment to disarmament and an unwillingness to develop WMDs in the country. In short, the membership of Iran is the result of its desire for disarmament and not vice versa. Otherwise, the experience of its war with Iraq and the silence of international community and organizations over the use of chemical weapons by Saddam Husain on Iranian cities have proved to Iran that international regimes are not helpful at the time of emergency, and they are "tools of western hegemony" (Patrikarakos 2012: 109).

Iran also accuses international regimes are not treating it fairly as a Non-Nuclear Weapons State. It argues that Western countries are violating its right to use nuclear power for the civilian purpose. It is discriminated from other nuclear-capable states like Argentina, Brazil, Japan, and Belgium which are not subjected to any international sanctions (Kibaroglu 2006: 2010). So, the historical experience of Iran shows that it is not benefitting, at least its leaders feel so, from international regimes. So, neoliberal institutionalism fails to explain why Iran is still a party to these regimes and why it is not withdrawing its membership and developing nuclear weapons.

Domestic Politics Model and Bureaucratic decision-making

Scott Sagan's domestic politics model has been used by many scholars to understand the nuclear policies of the Islamic Republic of Iran. This model "envisions nuclear weapons as political tools used to advance parochial domestic and bureaucratic interests" (Sagan, 1996:55). The works of Charles C. Mayer (2004), Mustafa Kibaroglu (2006), Oliver Schmidt (2008), Halit Mustafa Tagma and Ezgi Uzun (2012), and Michael Clarke (2013) analyze, and some of them even predict, nuclear weaponization of Iran using the domestic politics model. Halit Mustafa Tagma and Ezgi Uzun identify key actors in the nuclear decision-making of Iran as the Supreme Leader, the President, the Supreme National Security Council (SNSC), the Iranian Revolutionary Guards Corps (IRGC), and the Atomic Energy Organization of Iran (AEOI) (Tagma and Uzun (2012: 247). Mustafa Kibaroglu quotes Kayhan Barzegar, the director of the Institute of Middle East Strategic Studies in Iran, saying that "bureaucrats, scientists, and technicians who are directly involved in the nuclear projects are very concerned about halting the uranium enrichment process and stopping the nuclear projects because they are afraid of losing their jobs and prestige" (Kibaroglu 2006: 221).

However, in a political system such as Iran, an analysis without considering the power of the Supreme Leader would reach to the wrong conclusion. The Supreme Leader has an ultimate say over Iran's nuclear program just as the President has final say over the US foreign policy and nuclear program. According to Scott Sagan, three actors those commonly argue for nuclear weapons are state's nuclear energy establishment, important unit in professional military, and politician in states in which individual parties or the mass public strongly favour acquisition of nuclear weapons (Sagan, 1996: 64). However, in the case of Iran, as Oliver Schmidt says, "it is nearly impossible to find any proof for domestic actors actively lobbying for a nuclear-armed Iran" (Schmidt 2008: 60-61). As far as the military is concerned, Schmidt argues that since IRGC is subordinated to the Supreme Leader, it is hard to determine its significance in nuclear policies. According to Article 110 of the constitution, the Supreme Leader is the commander in chief of all armed forces (Buchta 2000: 46). Clarke classifies Iranian political leaders and elites into three factions: conservatives, reformists, and hardliners (Clarke 2013: 499). In another view, Iranian political leaders can be classified into four categories: moderate conservatives (led by Larijani), reformist (led by Khatami), moderate camp (led by Rouhani), and radical conservatives (led by Ahmedinejad). However, even Ahmedinejad, who

60

is usually categorized as a hard-line conservative, didn't argue for nuclear weapon program of Iran.

Hadian identifies four viewpoints among Iranian elites about nuclear weapons: First group, who are so-called "greens" and constitute 2% to 3% of the population, believes that Iran does not need nuclear weapons or the capability. The second group argues that "Iran is entitled to have the peaceful nuclear technology and it should not give up its right to peaceful applications of nuclear energy". The third group argues that Iran cannot trust the international community and hence, "needs to develop nuclear weapons capability, but not the weapons at this time". The fourth group, who are hardliners, "strongly argue for withdrawing from the NPT and developing nuclear weapons as soon as possible". According to Hadian "the first and fourth groups are unlikely under normal circumstances" (Quoted in Kibaroglu 2006: 220). Christopher L. Maurer quotes Gallup article of November 2013 arguing that when 68% of Iranian population supports peaceful nuclear program, only 34% supports the development of military capability (Maurer 2014: 64). In addition to that, even though some groups of politicians, bureaucrats, and military may have interests in developing nuclear weapons, they are not influential enough to overcome the stand of Supreme Leader and the majority of population.

As the Head of the Foreign Relations Committee of Iran's National Security Council for eight years (1997–2005), the co-author of the present book Mousavian is fully convinced that the strengthening and universalization of the Non-Proliferation Treaty and the establishment of a zone free from WMD in the Middle East and the elimination of such weapons altogether are compatible with Iran's security doctrine. Since the 1979 revolution, despite some rhetoric from radicals, the majority of Iran's prominent politicians have believed that the acquisition of a nuclear bomb would present a long-term threat to Iran's national interests, both regionally and internationally. Iran wants to be a modern nation, fully capable in advanced technologies. This ambition can be fulfilled only through normal relations with the international community. Acquisition of nuclear bombs would be counterproductive (Mousavian 2012b).

Individual Level Analysis

The psychological approach of Jacques E. C. Hymans and the mythmaking approach of Peter R. Lavoy are major individual approaches used to analyze the nuclear policies of Iran. Hymans's model explains the nuclear behaviour of states focusing on different modes of 'National Identity

Conception' (NIC). According to him, technologically nuclear-capable states would develop nuclear weapons when it is ruled by leaders with oppositional nationalist conception. In his words, "driven by fear and pride, oppositional nationalists develop a desire for nuclear weapons that goes beyond calculation, to self-expression" (Hymans 2006: 2).

However, this model is not enough to explain the absence of nuclear weapons in Iran. As it is mentioned earlier, Iran is a technologically capable state for developing nuclear weapons, and it also has been ruled by leaders with oppositional nationalist NIC. For example, Saira Khan has called Rafsanjani as an oppositional nationalist (Khan 2010: 16). Ahmedinejad was also a hardliner with pride in his nation and "us against them" (oppositional NIC) feeling about the relationship with western countries. However, Ahmedinejad did not order for nuclear weaponization of Iran.

According to mythmaking model of Peter R. Lavoy, two kinds of myths, security and insecurity myths, are propagated by leaders. If security myth prevails over insecurity myth, the country is likely to develop nuclear weapons. If insecurity myth prevails, the country will restrain from nuclear weaponization. Charles C. Mayer (2004) has argued that Iran has successfully employed both of these oppositional myths (Mayer 2004: 40). The security myth of Iran stresses the need for Iran's self-reliance, the threat from Israel and the US, the absence of support or nuclear umbrella of nuclear power, and the necessity of indigenous nuclear weapons for enhancing security and prestige of Iran. Nuclear insecurity mythmakers focus on the religious prohibition of nuclear weapons, impacts of weaponization on economic and foreign trade, the possibility of external attacks, and further regional proliferation as a consequence of weaponization. Out of these two myths, insecurity myth has a larger following since it is the formal position of Iran and "security myth makers are rarely heard outside of closed door" (Mayer 2004: 40, 42). However, these "rarely heard" voices are not enough for security mythmaking. So, the insecurity mythmaking is still dominant in Iran. As, Mayer (2004: 41) indicated, the ideology of Islam is a significant factor in this insecurity mythmaking.

Norms Focused Theories

The norms model of Scott Sagan explains nuclear behaviour of states as "weapons decisions are made because weapons acquisition, or restraint in weapons development, provides an important normative symbol of a state's modernity and identity" (Sagan, 1996: 55). Focusing on Iran's rejection of foreign intervention and domination, prestige over past civilization and greatness of a nation, scholars such as Oliver Schmidt (2008), Mustafa Kibaroglu (2006), David Patrikarakos (2012), and Michael Clarke (2013) predicted the possibility of nuclear weaponization. They explain nuclear weapons as a symbol of prestige, modernization, and greatness of Iran. Apart from just having them, nuclear weapons also work as a symbol of independence. Since the collapse of the Persian Empire and the decades-long domination of foreign powers, UK/Russia/US, the notion of independence has become the key factor for Iranians. Therefore, a nuclear weapon is not the prestige that Iranians want, instead, independence is the real national prestige and pride. After the revolution, the revolutionaries cancelled all nuclear programs related to domestic enrichment and heavy water. But when the US and EU decided to deprive Iran of its legitimate rights under NPT for peaceful nuclear technology, the Iranians went after self-sufficiency on nuclear technology, and it became a matter of national prestige and pride.

Oliver Schmidt (2008) and Michael L. Farmer (2005) have pointed to motivations of Iran for weaponization due to the decline of international norms on nuclear proliferation. They cite recognition of India as a nuclear power through India- US nuclear deal, beneficial proposals to North Korea, which was one member in the 'Axis of Evils' and lack of desire for disarmament from the Nuclear Weapon States as required by Article VI of the NPT as reasons for the decline of international norms on nuclear proliferation (Schmidt 2008: 66 and Farmer 2005: 39-40). The experience of the silence of the international community over the chemical attack by Iraq negatively affected the trust Iran had in international norms.

However, it is also important to focus on normative aspects which constrain Iran from weaponization. The Supreme Leader's religious fatwa claiming "that the production, stockpiling, and use of nuclear weapons are forbidden under Islam and that the Islamic Republic of Iran shall never acquire these weapons" is an important normative aspect to explain the absence of nuclear weaponization (Schmidt 2008: 68). Schmidt argues that if an anti-nuclear fatwa exists, "it would provide a very strong norm against nuclear weapons procurement" (Schmidt 2008: 76). So, the following section will explore the existence of this fatwa and its impacts in detail.

Role of Religion in Nuclear Policies

The Islamic Republic of Iran is a theocratic state, which recognizes the Twelver Shia branch of Islam as the state religion. About 98 percent of the population believes in Islam, and more than 90 percent belong to Shia branch. The Supreme Leader with the power of *Velayat-e-Faqih* (guardian jurist) is the most powerful political actor and final word in all political and religious matters (Altman 2009). He is commander in chief of the military, maker of foreign policy, and he has authority over the legislative, the executive, and the judiciary. He appoints members into key posts such as the commanders of the armed forces, the chief judge, and half of the 12 jurists of the Guardian Council. This Guardian Council has veto power to examine legislations passed by the parliament and nullify decisions which contradict to Islamic principles. As Institute for Science and International Security (ISIS) report indicated,

> "The Supreme Leader has an ultimate say over Iran's nuclear program. All major decisions on the nuclear issue, whether signing the Additional Protocol or suspending uranium enrichment, would require his approval" (ISIS Report 2013: 4).

Based on Article 176 of Iran's Constitution, major foreign and security issues are the responsibility of the Supreme National Security Council (SNSC), which plays very critical role in shaping Iran's foreign and security policies. US-Iran relations, the nuclear dossier, and regional issues, for example, are subjects that are part of the council's authority. Members of the council consist of the heads of the government's three branches, the chief of the Supreme Command Council of the armed forces, the chief commanders of the army and the Islamic Revolutionary Guard Corps, the head of the budget and planning, the ministers of foreign affairs, interior, intelligence and security, and another representative of the supreme leader. SNSC send the decisions for the supreme leader's confirmation. The decisions of SNSC require the approval of the Supreme Leader and fuqaha' of the Guardian Council, and he is the only one who can veto any decision made by SNSC. However, when the Leader approves, the decisions are considered national and should be implemented.

This introductory paragraph on the significance of religion in the state and society of Iran and power of the Supreme Leader is important to understand the impacts of fatwas of the Supreme Leader on nuclear decision-making.

Ayatollah Khomeini and WMDs

The first Supreme Leader of Iran, Ayatollah Ruhollah Khomeini, was an ideological critic of Weapons of Mass Destruction. It was reported that Khomeini had issued a fatwa against all kinds of Weapons of Mass Destruction during the time of War with Iraq (Sagan 2004: 87, Larssen 2011:50, Porter 2014). The strength of this religious position and its role in military policies were tested during the war with Iraq. The fatwa of Ayatollah Khomeini is cited as a reason for the absence of retaliation by Iran against the chemical attack on Iraq (Sagan 2004: 87, Porter 2014, Habibzadeh2014: 167, Mousavian 2012: 1, Mousavian 2013: 148, Mousavian and Afrasiabi 2012: 2, Collier 2003). Scott Sagan says that despite Iraq using chemical weapons against Iran, Ayatollah Khomeini opposed their acquisition or use by Iran, considering that it would be against the principles of Qur'an which prevent pollution of the atmosphere even during a holy war (Sagan 2004: 87). This experience of the war shows that the religious judgement of guardian jurist against chemical weapons overrides all other political and military considerations. The fatwa of Ayatollah Khomeini constrained Iranian force from developing WMDs and retaliating against Iraqi chemical attacks even though it caused a disadvantage to Iran in the war. The impact of the fatwa was not just the cancellation of chemical weapon program, but the fatwa also made it impossible for Iran to continue with the war (Porter 2014). Gareth Porter (2014) has published the experience of Mohsen Rafighdoost, who served as minister of IRGC throughout the wartime. Rafighdoost proposed to Ayatollah Khomeini in two separate meetings the Iran's planning of chemical and nuclear weapons and their necessity to retaliate against Iraqi chemical attack. However, Ayatollah Khomeini, who is also supreme commander of the military, rejected that plan since it is prohibited by Islam.

Ayatollah Khamenei on Nuclear Weapons

Ayatollah Ali Khamenei, who succeeded Ayatollah Khomeini as the Supreme Leader after his death in 1989, also stated against nuclear weapons on various occasions. The first fatwa of Ayatollah Khomeini on this issue was reported and publicized in 2004. However, Porter (2014) has argued that Ayatollah Khamenei issued an anti-nuclear fatwa (religious edict) in the mid-1990s as a response to a question on his religious opinion on nuclear weapons. Gawdat Bahgat also says that "Khamenei issued a *fatwa* in 1995 that considered all weapons of mass destruction as a great and unforgivable sin and declared them forbidden (*Haraam*)" (Bahgat 2003: 69). A fatwa is a legal ruling of a qualified jurisprudent on a given issue using recognized sources of Islamic jurisprudence and method of ijtihad (legal deduction) (Habibzadeh 2014: 154-155). Another definition of fatwa as per Shia jurisprudence is that "a fatwa is a religious legal opinion, ruling or decree concerning Islamic law issued, orally or in written form, by a prominent religious leader and expert in Islamic jurisprudence, based on four sources: the Koran, the practice of Prophet Muhammed and his successors, the power of reason and consensus" (Mousavian 2013: 147-148).

Seyed Hossein Mousavian (2013), Ayatollah Abolqasem Alidoost (2014), and Habibzadeh (2014) have quoted various statements of Ayatollah Khamenei against nuclear weapons. In 2004, Ayatollah Khamenei declared that "developing, producing or stockpiling nuclear weapons is forbidden under Islam" (Zakaria 2009). The official Iranian statement to IAEA in 2005 mentioned the fatwa of Ayatollah Khamenei against the production, stockpiling and use of nuclear weapons as forbidden under Islam (Mousavian 2013: 148). In 2008 Ayatollah Khamenei stated that "Iran has repeatedly declared that it is opposed to the production and use of nuclear weapons on fundamental religious grounds". In 2010 he declared that "according to our religious beliefs, use of these weapons of mass destruction is forbidden and 'haram'" (Habibzadeh (2014: 152). Addressing military commanders after a ship inauguration ceremony, he told them that, "Islam is opposed to nuclear weapons and that Tehran is not working to build them" (VOA News, 18-02-2010, Islam Today 20-02-2010). Ayatollah Khamenei repeated his stand on nuclear weapons in February 2012 during the meeting with officials of Atomic Energy Organization of Iran (AEOI) and nuclear scientists and in August 2012 at the 16th Non-Aligned Movement summit (Mousavian 2013: 148). Referring to the fatwa of Ayatollah Khamenei, Rolf Mowatt- Larssen (2011: 50) stated that "in light of various references, it would seem that Khamenei's fatwa is legitimate and absolute."

Supporting and Opposing Views

Various scholars expressed their supports or reservations on fatwa of Ayatollah Khamenei regarding nuclear weapons. Among early centuries' Shia scholars, Sheikh Tusi has prohibited spraying poison in the land of enemies (Habibzadeh 2014: 159, Alidoost 2014:3). This prohibition on poison is also applied to the use of non-conventional weapons. Among contemporary Shia scholars, many Grand Ayatollahs have the same fatwa, such as Grand Ayatollah Khoei, Grand Ayatollah Sistani, Grand Ayatollah Vahid Khorasani, Grand Ayatollah Hussein-Ali Montazeri, Grand Ayatollah late Mohammad Bagher Sadr, Grand Ayatollah Yusef Saanei, the Grand Marja of Shia Islam, Grand Ayatollah Sobhani, Grand Ayatollah Makarem Shirazi, and Grand Ayatollah Javadi Amoli. Since the rank of Majraal- Taqlid (who are Grand Ayatollah) is necessary for issuing a fatwa according to Shia rules, only the opinions of such Grand Ayatollahs are analyzed here. All such Shia Grand Ayatollahs, who are Marja Taghlid, unanimously believe that WMDs are haram (forbidden). Grand Ayatollah Yusef Saanei says:

> "There is complete consensus on this issue. It is self-evident in Islam that it is prohibited to have nuclear bombs. It is eternal law because the basic function of these weapons is to kill innocent people. This cannot be reversed." (Collier 2003).

Even though some scholars have expressed their reservations on this issue, they are not as much significant as some skeptics point out. Ayatollah Mohammed Taqi Mesbah Yezdi is the most quoted scholar permitting nuclear weapons under certain conditions. But, he is not in the position of Majra al-Taqlid to provide his independent opinion. It should also be noted that he has not issued any fatwa for developing nuclear weapons. The supporters of nuclear weaponization justify only possession of weapons for deterrence purposes. Tavakol Habibzadeh (2014: 155) says that,

> "Until now, other jurisprudents have not issued a fatwa permitting the use of WMD, but all the jurists who expressed their opinion in this regard issued a fatwa banning the use of nuclear weapons".

It is important to note that since the fatwa of the Supreme Leader binds on all political and legal systems of the state, even if few scholars issued fatwa supporting nuclear weapons, it will not affect the policies of the Iranian state. So, this religious position of Iran will not change until the Supreme Leader, SNSC, Parliament and majority of the Guardian Council change their views on nuclear weapons. Even though theoretically it is possible, it will take a long time. If the possibility of future change is considered as a factor to be a sceptic of existing law, it would be difficult to believe even laws passed by legislative assemblies. Because every law can be amended if the majority opposes it and it is more likely in secular laws than in religious judgements. Another scepticism is that some scholars could be lying for bluffing international community. However, as Fareed Zakaria pointed out,

> "It seems odd for a regime that derives its legitimacy from its fidelity to Islam to declare constantly that these weapons are un-Islamic if it intends to develop them. It would be far shrewder to stop reminding people of Khomeini's statements and stop issuing new fatwas against nukes" (Zakaria, 2009).

The pessimists of Ayatollah Khamenei's fatwa and nuclear program of Iran raise various doubts on the influence and role of this fatwa in actual nuclear decision-making. One argument is that Islam is a flexible normative foundation and can be interpreted at any point of time for legitimizing and for opposing nuclear weapons. Therefore, radical leaders could reinterpret it for supporting nuclear program (Khan 2010: 18, 19). The proponents of this argument cite opinions of scholars legitimizing nuclear weapons. However, all Shia Grand Ayatollas seated in Qom and Najaf, past and present, authorized to issue fatwa have unanimously confirmed that from the Shia Islamic point of view, all WMDs are haram. Apart from that, the number of supporters among other Shia scholars, who are not Grand Ayatollah nor Majra al-Taqlid, for nuclear weapons is very small, and opinion of such scholars is not relevant if it is contrary to the opinions of the Supreme Leader, other Grand Ayatollahs, and the Guardian Council. A change in the position of the Supreme Leader, SNSC, Parliament, administration, and Guardian Council is necessary to make a shift in nuclear policies of Iran. The change in the Islamic interpretation is not as easy as amending a secular law against nuclear weapons. The repeated statements of Ayatollah Khamenei and other Grand Ayatollahs have also produced a strong domestic consensus against nuclear weapons, and this consensus makes it difficult for ruling religious leaders to reverse this position (Collier 2003). The radical political leaders cannot legislate against the Islamic principles proposed by Guardian Council. According to Article 4 of the Iranian constitution, such laws are *null* and *void*.

The context of the fatwa and its durable relevance even out of that context is another concern for sceptics. Michael Eisenstadt says that "fatwas are not immutable, and no religious principle would prevent Khamenei from modifying or supplanting his initial fatwa if circumstances were to change" (Eisenstadt 2013: 2). This argument is an outcome of misunderstanding the role of context and basic principles in determining a fatwa. The fatwas which are based on some basic principles, such as the prohibition of indiscriminate mass killing, will not change because these basic principles are everlasting. Since the fatwa on nuclear weapons is based on such basic principles, its significance is also long-lasting (Mousavian 2013). Habibzadeh (2014: 156) pointed that the "nuclear fatwa is timeless, everlasting and having roots in centuries-old religious sources". In addition to that, the legitimacy of a Shia Marja is based on the credibility of his fatwas. Changing fatwa means losing legitimacy as a religious leader.

One criticism is that "the religious fatwa has no legislative basis and leaves significant room for manoeuvre" (Bowen and Moran 2014: 40). Such criticism arises due to the misunderstanding of Ayatollah Khamenei's fatwa as equal to the fatwa of any other qualified Muslim scholar. Fatwa of a normal qualified Muslim scholar binds only to those who follow him (*muqallideen*). At the same time, fatwa of the Supreme Leader binds state as a whole, and it has a legal status above mere legislation. (Porter 2014: 9). The fatwa of Ayatollah Khamenei overrides all other laws passed by legislators. If the legislative laws contradict with the fatwa of the Supreme Leader, the religious decree of the Supreme Leader will prevail. However, even though the fatwa is more powerful in Iranian legal system, Seyed Hossein Mousavian suggests that Iran could adopt legislation to outlaw development of any Weapons of Mass Destruction. According to Mousavian, this secularization of religious fatwa would help to remove ambiguity among the international community about the legitimacy of the fatwa. (Mousavian 2013: 157).

The absence of a written form of the fatwa is another source of doubt for critics. Ali M. Ansari (2013) argues that,

> "At present, such a fatwa – in any meaningful form - does not appear to exist. The concept of an "oral fatwa," in the context of Iran's nuclear program, is meaningless and for any such ruling to have any weight it should be detailed and substantiated, *and* ideally supported by a broad range of the senior clerical hierarchy in Iran today" (Ansari2013).

However, this concern is not significant considering the position of Ayatollah Khamenei and the publicity of his statements. As far as the legitimacy of a fatwa is concerned, it is not necessary to be issued in written form. It has been a practice since early times to issue oral fatwas, and it may be written down by those who heard it. The statements of Ayatollah Khamenei have also been reported by those who heard it. His statements against nuclear weapons have been published on his personal website (Khamenei 2012).

Above all, the full text of Ayatollah Khamenei's inaugural address delivered on 30 August 2012, at the 16th Non-Aligned Summit in Tehran is available on his website. He said:

> Honorable audience, international peace and security are among the critical issues of today's world and the elimination of catastrophic weapons of mass destruction is an urgent necessity and a universal demand. In today's world, security is a shared need where there is no room for discrimination. Those who stockpile their anti-human weapons in their arsenals do not have the right to declare themselves as standard-bearers of global security. Undoubtedly, this will not bring about security for themselves either. It is most unfortunate to

see that countries possessing the largest nuclear arsenals have no serious and genuine intention of removing these deadly weapons from their military doctrines and they still consider such weapons as an instrument that dispels threats and as an important standard that defines their political and international position. This conception needs to be completely rejected and condemned.

Nuclear weapons neither ensure security, nor do they consolidate political power, rather they are a threat to both security and political power. The events that took place in the 1990s showed that the possession of such weapons could not even safeguard a regime like the former Soviet Union. And today we see certain countries which are exposed to waves of deadly insecurity despite possessing atomic bombs.

The Islamic Republic of Iran considers the use of nuclear, chemical and similar weapons as a great and unforgivable sin. We proposed the idea of "Middle East free of nuclear weapons" and we are committed to it. This does not mean forgoing our right to peaceful use of nuclear power and production of nuclear fuel. On the basis of international laws, the peaceful use of nuclear energy is a right of every country. All should be able to employ this wholesome source of energy for various vital uses for the benefit of their country and people, without having to depend on others for exercising this right. Some Western countries, themselves possessing nuclear weapons and guilty of this illegal action, want to monopolize the production of nuclear fuel. Surreptitious moves are under way to consolidate a permanent monopoly over the production and sale of nuclear fuel in centres carrying an international label but in fact, within the control of a few Western countries. (Khamenei 2012a)

In addition to that, the declaration of Ayatollah Khomeini regarding WMDs was considered by Mohsen Rafighdoost, who was a minister of the IRGC throughout the war with Iraq, as a fatwa even though it was never written down or formalized because it was issued by "guardian jurist" whose statement will bind the entire government (Porter 2014: 6-7). Unlike the declaration of Ayatollah Khomeini, the fatwa of Ayatollah Khamenei on nuclear weapons has been formally written down by the government of Iran. For example, in a letter to IAEA Iranian government quoted fatwa of Ayatollah Khamenei stating that "the production, stockpiling and use of nuclear weapons are forbidden under Islam and that the Islamic Republic of Iran shall never acquire these Weapons" (Habibzadeh 2013: 164).

Mohammad Khazaee, the Iranian Ambassador to the UN, told Mousavian, one of the co-authors of this book, that he submitted the fatwa of Ayatollah Khamenei through an official note to UN Secretary-General during the second term of Ahmadinejad's presidency to be registered at the UN as official and religious position of Iran on ban of all WMDs.

Another worry comes from Mehdi Khalaji (2013: 14) when he argues that Ayatollah Khamenei's prohibition has shifted gradually "from a denial of the practical utility of nuclear weapons to a focus on Islamic prohibitions against their use". Khalaji says that the statement of the Supreme Leader in recent years focused only on the use of weapons, and it may be due to the Leader's view that "creating and storing such weapons will be sufficient to change the power equation in the region" (Khalaji 2013: 14). However, such criticism is baseless since Ayatollah Khamenei also stated against possession of nuclear weapons in 2012 in his statement during the meeting with officials of Atomic Energy Organization of Iran (AEOI) and nuclear scientists (Mousavian 2013: 149).

Importance of the Fatwa and Religion in Nuclear Policies

Considering the power of the Supreme Leader in the hierarchy of power structure, his religious decree has a decisive influence in the nuclear decision-making of Iran. His position as 'guardian jurist' strengthens the religious and political importance of his fatwa. Religiously, his fatwa is binding not only on his followers (*Muqallideen*) but also on all people. Even if other *mujtahids* have a different fatwa regarding this issue that would not be valid (Habibzadeh 2014: 169). Therefore, the position and power of Ayatollah Khamenei ensure the long-lasting of this religious position of Iran without being challenged by other scholars, though most of the scholars are in support of Ayatollah Khamenei.

Politically, "according to Article 57 of the constitution, the Supreme Leader has ultimate authority over three branches of government, and his command is binding" (Mousavian 2013: 154). His fatwa binds the whole state mechanism and overrides all other military and political considerations. According to the Iranian constitution (Article 4), any law contrary to Islamic principles is null and void. So, even if the assembly or executive gives permission for developing nuclear weapons, that decision would not be constitutionally valid. In addition to that, all legislation adopted by the assembly must be referred to Guardian Council and that council has the power to veto any law if it contradicts with Islamic principles. It means that the religious decree and principles are stronger than a mere legislation.

The fatwa of Ayatollah Khamenei and other Shia Grand Ayatollah sitting in Iran, Iraq, Lebanon, and Pakistan could create a strong domestic public opinion and norms against nuclear weapons inside Iran and beyond. Even those sceptical about the existence of such fatwa agree that "if an anti-nuclear fatwa really exists, it would provide a very strong norm against nuclear weapons procurement" (Schmidt 2008: 76). Ayatollah Sadeq Amoli Larijani, head of the Iranian judiciary, stated that "the fatwa that the Supreme Leader has issued is the best guarantee that Iran will never seek to produce nuclear weapons" (Bowen and Moran 2014: 40). Nevertheless, Iranian foreign minister Ali Akbar Salehi has declared Iran's willingness to transform the fatwa into a legally binding official document in the UN (Mousavian 2013: 147).

Conclusion

The Islamic revolution of 1979 in Iran affected the nature and motivation of its nuclear program. Islamic and anti-colonialist ideologies became shaping factors in post-revolutionary Iran's nuclear program. The theocratic political system provides the religion and the Supreme Leader decisive power. Therefore, the religious principles and views of the Supreme Leader and Guardian Council are significant factors in determining nuclear policies of Iran especially when all Shia Grand Ayatollahs have the same fatwa. The religious principles override other material interests of the Iranian state. The constitution of Iran strengthens the power of these religious authorities by providing them veto power over legislations and decisions passed by legislators and executives. Since the Supreme Leader is the commander in chief, his opinion is binding on the military as well. In short, as a theocratic state, the views of religious authorities have a significant role in military and security policies, including nuclear weapons, of the Iranian state.

As far as nuclear weapons are concerned, the Supreme Leader Ayatollah Khamenei has repeatedly issued his statements at many national and international venues, against nuclear weapons. According to him and many other Ayatollahs, the possession and use of nuclear weapons are prohibited by Islam. The position of the Supreme Leader as guardian jurist strengthens his fatwa and makes it binding on all state departments and on all people. Though very few Islamic scholars have criticized Ayatollah Khamenei's fatwa, his power ensures that these fatwas are left unchallenged. Since the legitimacy of the government among people is rooted in religious identity and the norms against nuclear weapons are popularized through the repeated statements of authorities, it is hard for Iran to reverse its position.

Therefore, the development of nuclear weapons by Iran is unlikely in the near future.

Theoretically, this influence of religion can be analyzed by using different theories. However, rationalist approaches are not sufficient to explain Iranian security and nuclear behavior. Since domestic politics and power structure are significant factors, Iran cannot be considered as a "black-box". Even though the approaches which focus on domestic political structure can deal with it, their framework is to be broadened to include theocratic governments rather than binaries democratic and autocratic governments. It is necessary to understand the influence of religious principles on government policies. Even though mainstream constructivist theories, which consider norms as a significant factor, have not given much attention to the role of religion in shaping these norms, the norm model of Scott Sagan can explain the influence of religion. Individual-level analyses have to take into account ideological and normative motivations of the individual. The mythmaking theory of Lavoy can accommodate this religious influence in nuclear policies of Iran. However, it has to consider religion as a source of motivation to leaders and mythmakers rather than a tool to spreading their myths.

Chapter IV

Role of Religion in Nuclear Policies of Pakistan

Introduction

Pakistan is the sole Muslim-majority country which has developed nuclear weapons and claimed Nuclear Weapons States status. Zulfikar Ali Bhutto, who diverted Pakistan's nuclear program into military purpose, argued that "the Christian, Jewish and Hindu civilizations have this capability; the communist powers also possess it; only the Islamic civilization was without it, but that position was about to change" (Bhutto 1979: 151). After this statement, the academic circle has been discussing the role of Pakistan's nuclear weapons as an 'Islamic bomb" and its implication in Arab-Israel conflicts. D. K. Palit and P. K. S. Namboodiri (1979) analyzed the Islamic nature of Pakistan's bomb and support of other Muslim countries for this cause. Herbert Krosney and Steven Weisman (1981) also analyzed the "Islamic bomb" and its implication in Arab-Israel conflicts. They examined pan-Islamic nature of Pakistan's nuclear program and religious motivation for Pakistan and other Muslim states to develop a nuclear bomb by an Islamic country.

At the same time, as described in the previous chapter, Iranian Supreme Leader Ayatollah Khomeini opposed nuclear weapons and contended that nuclear weapons are prohibited by Islam. Pakistan's leaders such as Benazir Bhutto, Nawaz Sharif, and Pervez Musharraf rejected any pan-Islamic aspect for Pakistan's nuclear weapons. According to them, Pakistan's nuclear weapons are for national security and for deterring attack from India (Hashmi 2004: 341). These arguments and counter-arguments raise different questions. Were the nuclear weapons of Pakistan motivated by religious cause?; Why didn't religion constrain nuclear weaponization in

Pakistan as the same religion did it in Iran?; What are the motivations of Pakistani leaders like Z. A. Bhutto to present its nuclear program as an 'Islamic bomb' and why do the leaders of the same state later reject such a label calling it as propaganda by pro-Indian and pro-Israel lobby?; What is the role of Pakistan's nuclear weapons in Arab-Israel conflict and what are the possibilities of transferring technology to other Muslim countries?

This case study of Pakistan tries to answer these questions through historical and theoretical analysis. The first section goes through the history of Pakistan's nuclear program and explains the context in which Pakistan decided to shift the program into military purpose. The following section analyzes the explanations of different theories about nuclear weapons of Pakistan and examines different factors which caused the weaponization. The third section investigates the role of religion in the nuclear decision-making of Pakistan through the prisms of various theories. The last section discusses the validity of the label "Islamic bomb." This section analyzes arguments for and against this label. It also explains the possibility of extended deterrence and transfer of nuclear technology to other states for the cause of Islam.

History of Nuclear Program in Pakistan

The nuclear program of Pakistan was started in 1954 after seven years of independence. Due to many political, economic and technological reasons, it was late compared to the nuclear program of India which is its neighbouring rival country and got independence about the same time. As B. Chakma (2002: 874) and Z. Khan (2015: 21) outlined, the reasons were absence of domestic political environment in Pakistan to launch a nuclear program; lack of awareness about civilian use of nuclear technology in fields such as agriculture, energy, and medicine; lack of nuclear 'enthusiast' political leaders and scientists match to Jawaharlal Nehru and H. J. Bhabha; and lack of industrial infrastructure as a potential to nuclear research program. R. K. Abbasi (2010: 67, 2012: 87) claims that Pakistan possessed none of the variables which were necessary for initiating a nuclear program soon after its independence. Most of the industries and scientific laboratories of British India were located in the Indian side of the border. Therefore, Pakistan had to focus on basic industries and the political and economic survival of the state. In this context, only very few military leaders and scientists thought it might ever be possible to develop a nuclear bomb (Hoodbhoy and Mian 2014: 1127).

Pakistan established Atomic Energy Committee (AEC) inspired by the 'Atom for Peace' proposal of the US. Following the recommendation of the Committee, Atomic Energy Council was set up in 1956, and it was later renamed as Pakistan Atomic Energy Commission (PAEC). The activities of PAEC in its first decade were oriented to the peaceful use of nuclear energy. At least until the early 1960s, Pakistan had no serious plan for military use of nuclear weapons. According to B. Chakma, there is no evidence of Pakistani consideration to use nuclear power for weaponization (Chakma 2002: 878). There were a lot of factors behind this non-weaponization. The political leaders were not interested in nuclear weapons. Even though Pakistan faces a security threat from India since its birth itself, its leaders preferred to counter this threat by strengthening the conventional power and through external balancing by joining with the US-led western military alliance such as the Southeast Asia Treaty Organization (SEATO), and the Central Treaty Organization (CENTO).

The development of civilian nuclear infrastructure also was very slow. Financial constraints, bureaucratic troubles in PAEC and lack of skilled manpower were significant hurdles in the progress of the nuclear program. Nazir Ahmed, the PAEC Chairman, complained that "the procurement of nuclear reactors was being delayed for 'non-technical' (financial) reasons." Nazir believed that bureaucrats are the main constraint to any project which he plans to implement. However, R. K. Abbasi (2010: 69, 2012: 89) points out the lack of political skill in Nazir to mobilize political support and resources for his project as an important barrier.

Zulfikar Ali Bhutto, who took charge of the Ministry of Fuel, Power, and Natural Resources in 1958 made significant changes in PAEC. He appointed I. H. Usmani as head of PAEC replacing Nazir Ahmed. Usmani accelerated the nuclear progress by establishing training and research facilities and seeking international recognition and cooperation for the peaceful nuclear program. After 1960, Ayub Government allocated more funds to the nuclear sector. Like many other developing countries, Pakistan also sent its scientists to developed countries such as the US for training. However, since many of them did not return to home country after their training, lack of skilled human resources continued as an obstacle for Pakistan. Despite the existence of these impediments, the nuclear program of Pakistan 'gradually acquired momentum' in the early 1960s (Chakma 2002:876).

Although President Ayub continued in his stand against nuclear weapons, few occurrences of 1960s created debate within Pakistan regarding possibilities and necessities of nuclear weapons. After the nuclear test of

China in 1964, hawkish wing led by Z. A. Bhutto, who was then foreign minister of Pakistan, expected that India would also follow China and develop its own nuclear weapons. The statement of H. J. Bhabha, the Chairman of the Indian Atomic Energy Commission, in 1964 regarding the low cost of nuclear weapon development and capability of India "to make a bomb in 18 months if it wanted to" were heard in Pakistan with high concern. Z. A. Bhutto had warned Ayub of the possibilities of nuclear weaponization of India in early 1963. However, Ayub opposed the nuclear weaponization, and he viewed Pakistan had no economic, political, and military need for them (R. K. Abbasi 2010:81, 2012: 103). The expectation was that Pakistan can buy finished nuclear weapons from its western allies if it is attacked by India. It is reflected in the statement of Ayub when he said that "if India went nuclear, we would buy a (nuclear) weapon off the shelf somewhere" (Chakma 2002: 878-879). Ayub added that "what do we need a bomb for? Pakistan is a poor country.....We cannot afford it.... We should put money into schools, maybe hospitals and industry" (Z. Khan 2015: 22). Theoretically, this statement of a military ruler is contradictory to the domestic politics-oriented theory which expects military leaders to be more supportive of the development of nuclear weapons.

The war with India in 1965 was another incident which triggered debate for nuclear weapons. The war proved the fact that the conventional weapons of Pakistan could not counter much superior force of India, and it strengthened the hawkish argument for nuclear weaponization. The competing groups debated over 'the desirability, feasibility, and utility of nuclear arms acquisition' (H. Abbas 2008: 95). Z. A. Bhutto declared that "even if Pakistanis have to eat grass, we will make the bomb." The response of Washington during the war also made Pakistan disappointed. Despite an alliance between Pakistan and the US as SEATO and CENTO, the US did not help Pakistan and moreover imposed an arms embargo on both India and Pakistan. Betrayed over a stand of the US during the war, Pakistan withdrew from SEATO and strengthened its relationship with China.

The expectation of nuclear help from the US declined after 1965 war and the underdeveloped nuclear infrastructure did not provide any scope for diverting nuclear programs for military purposes. Therefore, President Ayub preferred strengthening the conventional military forces of Pakistan (Hoodbhoy and Mian 2014: 1127). He rejected Bhutto's plan for 3,000 reprocessing plants. Failing to win President's support for nuclear weapons, Bhutto resigned from the foreign ministry in 1966 and started to mobilize public support for this cause. However, Ayub kept the door for nuclear weaponization in future open by refusing to sign the Nuclear Proliferation Treaty (NPT). Pakistan had played a constructive role during the NPT negotiations expecting that it would hinder India's nuclear weaponization. When Pakistan understood that India would not sign the NPT, it also followed the suit. It was a clear manifestation of an Indo-centric aspect in nuclear policies of Pakistan (Chakma 2002:883). Zafar Khan argues that the 'India factor' which was developed in the statement of Bhutto and became official in Pakistan's nuclear policies option indicated the primary rationale of Pakistan's nuclear program after Bhutto came into power in 1971 (Z. Khan 2015: 24).

The focus of Pakistan's nuclear program shifted into military purpose in the 1970s. Three events influenced this change to nuclear weaponization. First of all, the humiliation of defeat and disintegration of Pakistan in the war with India in 1971 reinforced the view that only nuclear weapons can deter India. Even though the creation of Bangladesh was the result of civil war responding to discrimination and unpopular policies of leaders belonging to West Pakistan, nuclear hawks calculated that if Pakistan had nuclear weapons, India would not have dared to involve in the war. A. Q. Khan explicitly stated in 2011 that "if we had had nuclear capability before 1971, we would not have lost half of our country—present-day Bangladesh— after disgraceful defeat" (Hoodbhoy and Mian 2014: 1129). B. Chakma identifies various implications of the war in the military policies of Pakistan. First, this war proved again that the conventional power of Pakistan is far inferior to India. Second, Pakistan lost not only its eastern wing, but also its strategic geographical advantage to fight a two-front war with India in both East and West side. The war strengthened the fear of Pakistan that India is destroying Pakistan to establish Akhand Bharat (undivided India) by weakening the two-nation theory (Chakma 2002: 885). The 1971 war also validated realistic suggestions for internal balancing and self-help. Despite an alliance with the US, Washington refused to assist Pakistan in its fight with India. Besides, China did not aid Pakistani militarily except some verbal threats against India. This experience led the Pakistani leaders to conclude that only internal balance through developing nuclear weapons can deter India and ensure the survival of Pakistan.

The election of Zulfikar Ali Bhutto to the post of President was the breakthrough in the nuclear weaponization of Pakistan. B. Chakma identified four important steps taken by Bhutto soon after assuming presidency facilitating nuclear weapon development. First, Bhutto appointed himself in charge of the Division of Nuclear Affairs. Second, the activities of the PAEC were rendered with more secrecy. Third, Munir Ahmed Khan, who was the advocate of nuclear force, was appointed as the Chairman of the PAEC replacing I. H. Usmani, who was no supporter of nuclear weaponization. The last, but not least, was an emergency meeting of nuclear scientists at Multan where they assured him, they could make an atomic bomb within three years if they get enough finance and facilities (Chakma 2002: 886-887). Considering the role of Bhutto in the development of both civilian and military nuclear development, he has been called the political father of Pakistan's nuclear program (Hilali 2002:7).

'Peaceful Nuclear Explosion' (PNE) of India in 1974 was another decisive event in determining and accelerating nuclear weaponization of Pakistan. The media, general public or government of Pakistan did not believe Indian claim that the explosion was a peaceful one. For them, the test posed a grave security threat and 'blackmail' to Pakistan. According to Hoodbhoy and Mian (2014: 1128), the explosion made 'the quest for bomb become a scramble' in the country. H. Abbas says that "the Indian test was the tipping point that altered the earlier 'capability decision' into a 'proliferation decision'" (Abbas 2008: 103). Following the test, the cabinet defence committee meeting of Bhutto's government officially approved building a nuclear bomb. Pakistan accelerated the program by employing both plutonium and uranium routes of weaponization.

Zia-ul-Haq, who succeeded Z. A. Bhutto through a military coup, continued nuclear weaponization. Zia pursued 'nuclear ambiguity' as a strategic posture without admitting or denying the military nuclear program. At the initial years of Zia's rule, the nuclear program has faced difficulties due to the pressure of US-led western countries to abandon nuclear weaponization. France cancelled its Nuclear Reprocessing Plant Agreement. Since the Western countries tightened the export of nuclear technology and materials after PNE of India, Pakistan faced difficulty in getting them from international markets.

However, the shift in international politics and change in priorities of the US after USSR's occupation of Afghanistan changed the dynamics. To get rid of USSR from Afghanistan was the top priority for the US than non-proliferation in South Asia. Since the cooperation of Pakistan was necessary for fighting against the USSR, the US suspended application of uranium enrichment sanction on Pakistan for six years and increased economic and military assistance of $3.2 billion (Cirincione, et al. 2005: 244). The Regan administration bypassed Solarz Amendments and Pressler Amendments for aiding Pakistan. According to this Pakistan-specific amendment, the economic and military sanctions would be implemented against Pakistan unless the President of the US certifies that Pakistan is not developing nuclear weapons. However, even after Pakistan officials claimed that it had acquired nuclear capability, and intelligence reports confirmed it, Regan's and Bush's administrations certified that Pakistan has no nuclear weapon program (Kerr and Nikitin 2016:5, Ahmed 1999: 187). The Soviet invasion of Afghanistan also increased Pakistan's security threat.

Zia-ul-Haq skillfully exploited the new international situation and accelerated nuclear weapon development. A.Q. Khan claimed in December 1984 that "KRL (Khan Research Laboratory) was in a position to detonate a nuclear device on a week notice" (Quoted in Abbas 2008: 110). Feroz Hassan Khan argues that Pakistan developed the nuclear device by the end of 1984s. However, "it was a large bomb and could be delivered by only a C-130 cargo aircraft with no assurance of delivery accuracy" (F. H. Khan 2012: 189). A. Q. Khan claimed in 1987 that "what the CIA has been saying about our possessing the bomb is correct ...... we shall use the bomb if our existence is threatened" (Chakma 2002: 899). National Intelligence Estimate (NIE) of 1991 assessed that Pakistan has viable nuclear weapon design and components which can be assembled into a bomb on short notice. Pakistan attained such a capability "by the end of the 1980s" (Kerr and Nikitin 2016:5). In short, even though there is no consensus on the exact time when Pakistan developed a nuclear device, it attained such a capability by 1980s.

However, the withdrawal of the USSR from Afghanistan and the end of Cold War reduced the geopolitical importance of Pakistan. Washington restarted its sanctions on Pakistan under Pressler Amendment. President Bush refused to certify that Pakistan did not possess nuclear weapons. The change in the attitude of the US made Pakistan feel 'betrayed' and 'abandoned'. This outcome caused a reassessment of security impact and risk of the dependence on an external military power and provoked Pakistan to expedite nuclear weaponization (Chakma 2002: 897).

Meanwhile, the ambiguity in nuclear policies and reports on weaponization had worked as an opaque deterrent between India and Pakistan. It is argued that it was nuclear capability which played a role as a deterrent in Brasstacks military crisis of 1986-87 between India and Pakistan. It also confirmed the belief of many Pakistanis that only nuclear weapons can deter a conventional attack of India (Chakma 2002: 900). The Kashmir Crisis in 1990 was another boost to the nuclear program of Pakistan when its leaders realized that only presence and threat of nuclear weapons deter

India (Hoodbhoy 1995: 8, Chakma 2002: 906, Z. Khan 2015: 30). The previous experience of isolation and lack of support from the US during the time of crisis and confidence in deterrence capability of nuclear weapons led Pakistan Army to reject the trade-off between nuclear deterrence and conventional weapons offer of the US such as F-16. Army considered the F-16s were not alternative for nuclear capability (Chakma 2002: 9007).

Since Pakistan acquired nuclear capability by 1980s, the tit-for-tat response to Indian test of 1998 was not a surprise. Even in 1996, after a report on Indian preparation for the nuclear test, John Deutch, then the Director General of Central Intelligence Agency (CIA), has said that "we have judged that if India should test, Pakistan would follow" (Cirincione, et al. 2005: 241). Pakistan's decision not to sign the Nuclear Proliferation Treaty (NPT) renewal of 1995 and the Comprehensive Test Ban Treaty (CTBT) of 1996 following the same decisions of India was an indication of Indo-centric nuclear policies of Pakistan and of its unwillingness to close the door for a test of nuclear weapons. Therefore, regarding the nuclear test of Pakistan on 28 and 30 May 1998 Zafar Khan concluded that "Pakistan's quest to acquire nuclear capability was neither miracle nor an overnight development" (Z. Khan 2015: 33).

Theoretical Explanations for Nuclear Weaponization of Pakistan

There are a lot of factors and motives which caused Pakistan's nuclear weaponization. Most of them are related to Pakistani perspectives about India either directly or indirectly. Therefore, most of the theories follow Indo-centric explanation for Pakistan's nuclear development. Even though "there is a consensus among academics that Pakistan's decision to conduct tests was a product of Indian-generated security imperatives" (H. Abbas 2008: 85), this chapter also analyzes other theoretical explanations to get broader understanding of the issue and to investigate the role of religion in the nuclear decision-making of Pakistan. Therefore, this section evaluates the nuclear program by using international, national, and individual levels focused theories.

The Bomb for Security

The security-focused realism explanation is still a dominant approach in the nuclear proliferation theories especially to the nuclear program of Pakistan. Even the scholars belonging to other theoretical frameworks consider security concern as the central element. For example, Feroz Hassan Khan (2012) mixes realism with strategic culture. Peter R. Lavoy (2006) also considers insecurity as the main component of the strategic culture of Pakistan. B. Chakma (2002: 873) argues that ".... ultimately it was a concern for national security, and survival played the critical role in turning it to a military-oriented project". R. K Abbasi quotes Hassan Askari contending that "realism rules the region and security remains the main driver for Pakistan's nuclear weapons development" (R. K. Abbasi 2010:4).

The basis of the insecurity is rooted in geographical, political, historical, and cultural dynamics of Pakistan. Geographically, Pakistan located in the middle of three powerful neighbors, i.e., China, India and Iran. Even though all of these neighbors have memories of golden past empire and ambition to be great powers, threat from India is comparatively at a higher level. The antipathy between India and Pakistan rooted in the origin of both countries. Pakistanis still fear that India rejects the basic idea of 'two-nation theory" which paved foundation for the creation of Pakistan (Lavoy 2006:9). Subsequent events such as the war of 1971 war and the creation of Bangladesh reinforced this fear. Since the concentrated population centers are near to Indian border, it increases the geographical vulnerability of Pakistan.

Pakistan was born in an insecure situation and experienced three wars with India in 1947-48, 1965 and 1971. These wars and the separation of Bangladesh exacerbated this fear (Abraham 2009: 5-6). The lesson of these wars was that Pakistan could not win against India using only conventional weapons. India has superiority over Pakistan in all forms of national power such geographical, economic, and military strength. According to realism, Pakistan could either balance against India or bandwagon with it. The bandwagon option requires accepting the dominance of India and being satisfied with status-quo. Pakistan is not ready to accept the status-quo in South Asia with the dominance of India (F. H. Khan 2005: 7-8). The balancing strategy may be external balancing through alliance building or internal balancing through strengthening national economic and military power. In first two decades of independence Pakistan, it preferred external balancing and joined US-led military alliance such as SEATO and CENTO. However, validating the argument of Mearsheimer, "alliances are only temporary marriages of convenience: today's alliance partner might be tomorrow's enemy" (Mearsheimer 2001: 31), the relation with the US was not reliable and was not much useful in the time of crisis. Even though the goal of Pakistan by allying with the US was to counter India, the goal of the US was to fight against the USSR. So, Pakistan got the support of the US only when the later needed the support of Pakistan to prevent Soviet influence in the region. In Feroz Khan's words,

"U.S. security guarantees, so enticing to Pakistan, were found to have no utility when Pakistan faced Indian forces in 1965 and 1971. Pakistan drifted from the "most allied ally" in the 1950s and 1980s to the most sanctioned ally in the 1990s, to the "most suspected ally" from 2001 onwards." (F. H. Khan 2005: 8).

The experience of 1971-war and lack of enough support from the US and China taught Pakistan the significance of self-help and strategy of internal balancing for survival (A. Z. Hilali 2002: 4). Due to economic constraints, it was not easy to invest a huge amount in conventional weapons to deter India. In the words of Bhutto, "a non-industrialized country, without even the basis of heavy industry, cannot depend entirely on the traditional defense system of a small, though highly efficient, armed force equipped with conventional weapons" (Bhutto 1967: 129). The nuclear weapon was the only option to ensure security and to avoid the repetition of 1971-experience. Therefore, in the perspective of realism, as Sumit Ganguly argued,

"the core aim of nuclear weapons of Pakistan was to prevent a repetition of 1971 and to deter an Indian attack that might reduce Pakistan's size even further, or perhaps even put the country out of existence entirely" (Quoted in H. Abbas 2008: 84).

The Bomb for Prestige

The "prestige" and "status" are significant motives for states in developing nuclear weapons. According to prestige-oriented theories, nuclear weapon is seen as a symbol of modernity and scientific development and states develop nuclear weapons to improve its international status. The prestige variable also has a role in the nuclear weaponization of Pakistan (Chakma 2002: 911). Pervez Hoodbhoy has considered prestige factor as a significant one to understand the nuclear behaviour of Pakistan. Hoodbhoy and Mian (2014: 1129) contend that the Urdu equivalent of the term "deterrence," i.e., *sad-e-jaeehat* is unfamiliar to most of the Pakistani people. For the common people, the nuclear program was part of *izzath* (pride) of the nation.

The pride factor in Pakistan has three aspects. First, nuclear weapons provide a sense of Pakistan's parity with India. Since the partition, Pakistan was keen to get the equal status of India and want to free from all forms of inferiority. This status was threatened when Pakistan was defeated against India in three wars. Therefore, the development of nuclear weapons was part of *izzath* of the nation, and it was more than just seeking security.

Second, nuclear weapons were proof of defiance of the pressure from great powers such as the US. The nuclear development after the decades-long pressure of the US provided a sense of pride to common people. Third, Pakistan considered nuclear weapon as a factor to improve its status among Muslim countries. As the first and sole Muslim country with nuclear weapon capability, Pakistan saw in the bomb a chance to claim leadership of the Muslim states.

Domestic Level Analyses

The domestic level analyses focus on factors such as bureaucratic and political parochial interests, and domestic strategic culture. Samina Ahmed (1999) has explained Pakistan's nuclear development and turning points by focusing on the parochial interests of military and bureaucrats. According to her, political leaders have only marginal role in the final test of 1998 (Ahmed 1999: 178). Even though the policies of the military are directed by security threat from India, the interpretation of security had served the parochial interest of the military. The continuous hostile relation with India legitimized a massive amount of defense budget and military rule in Pakistan (Ahmed 1999: 179). S. H. Qazi in his review of the book *Eating Grass* of F. H. Khan has argued that reading between the lines of the text indicates that it was domestic politics, rather than national security, that motivated nuclear weaponization (Qazi 2013: 90). He points to the coalition between Z. A. Bhutto and PAEC scientists along with some foreign ministry bureaucrats who have a similar view. B. Chakma expresses another view when he contends that the intra-bureaucratic politics have played a role in nuclear policies in its early stages. From the 1970s onwards, it was Indo-centric strategy which motivated the nuclear program (Chakma 2002: 911-912).

Pakistan's nuclear weaponization has also been explained based on its strategic culture. The works of Feroz H. Khan (2005 and 2012) are excellent examples of it. Strategic culture has been defined as "a collectivity of the beliefs, norms, values, and historical experiences of the dominant elite in a polity that influences their understanding and interpretation of security issues and environment, and shapes their responses to these" (Rizvi 2002: 305). According to F. H. Khan, the strategic culture of Pakistan is shaped by its historical experience with India including dismemberment of Pakistan in 1971 and the nuclear test of India in 1974. Then, for F. H. Khan, the components of strategic culture are also Indo-centric security-related factors. Image of Pakistan to itself and pride in its history are significant factors in shaping its strategic culture (F. H. Khan 2005:4). These aspects of strategic culture motivate Pakistan to compete with its three powerful neighbors with a history of great civilization, i.e., Chinese, Indian and Persian. It tries to defend all forms of domination of neighbors and to keep its separate identity as a Muslim nation. However, it is still afraid of the attempt of India to destroy the two-nation theory which is the basic idea of Pakistan. The acceleration of nuclear weaponization after 1971 war as a deterrence against India was to protect this image and distinguish the identity of the nation.

However, Lavoy indicated the limitation of 'strategic culture' as a variable to explain Pakistan's nuclear program. Because, since the dominant security organization in the country is the military force which is very conservative and pro-western the approach of strategic culture would lead into the conclusion that Pakistan would continue to rely on conventional weapons and strategic relationship with the US (Lavoy 2006:19). So, according to Lavoy, it is difficult to explain the shift to nuclear weaponization through the prism of strategic culture. For Lavoy, "mythmakers redefine and transform the strategic culture in line with their own strategic preferences and their understanding of the area" (R. K. Abbasi 2010: 3).

Individual level Analyses

Individual level oriented theories investigate the role of the nationalist feeling of the leaders (According to Hyman National Identity Conception-NICs) in nuclear decision-making and the role of the individuals as mythmakers. These approaches mainly emphasize the contributions of Zulfikar Ali Bhutto and Zia-ul-Haq in the development of nuclear weapons. The psychological approach of Hymans explains how the worldview, motives, and nationalist feeling of leaders shaped nuclear weaponization. In the case of Pakistan, it was Z. A. Bhutto, who has been called as 'political father of Pakistan's nuclear program,' who diverted Pakistan's nuclear program into military purpose. According to the classification of Hymans, Z. A. Bhutto can be termed as "oppositional nationalist". However, he could not produce the bomb due to technological underdevelopment of Pakistan.

The mythmaking approach of Peter R. Lavoy also judges the contribution of Z. A. Bhutto as a significant factor in Pakistan's nuclear weaponization. According to Lavoy, mythmaking approach "performs better" in the explanation of Pakistan's nuclear weaponization (Lavoy 2006: 19). Just like other countries, there were both security and insecurity mythmakers in Pakistan. During presidency of Ayub Khan, the position of insecurity mythmakers led by the president prevailed over security mythmakers led by the then Foreign Minister Z. A. Bhutto. However, Z. A. Bhutto continuously popularized his position. After the defeat of Pakistan in 1971 war and the emergence of Bhutto as the president, the security myth was gained prominence. Feroz H. Khan points into three aspects in this strategic mythmaking. First, only nuclear weapons can guarantee survival against Indian military supremacy. Second, the international opposition to Pakistan's nuclear bomb is due to its Muslim population. Third, India, Israel, and the US may ally and use force to stop nuclear program of Pakistan (F. H. Khan 2012: 6). Z. A. Bhutto was successful in popularizing his view. The support for nuclear weapons by a minority before 1971 became a national consensus by 1974 (F. H. Khan 2012: 6). Since the position of Bhutto on linking nuclear weapons with national security and honour was accepted by the Pakistani people, it was difficult to subsequent governments or leaders to compromise on it (Hilali 2002:9).

Theoretically, the nuclear weaponization of Pakistan can be explained through the frameworks of different theories. Out of all these theories, the security-oriented realism is dominant. Other theorists, who focused on individual domestic levels, also have considered the importance of security threat from India as a significant factor. However, the understanding of other variables such as prestige, domestic politics, strategic culture, and individuals provide a comprehensive picture of the nuclear policies of Pakistan. Even though security is the dominant factor, other factors also have a crucial role in interpreting nature of security threat and counter-strategies and to transforming this threat to nuclear weaponization. In short, as B. Chakma (2002: 911) contended Pakistan's nuclear weaponization was "necessarily a multi-causal phenomenon."

Role of Religion in Nuclear Policies of Pakistan

Role of Religion in Pakistan

The analysis of the role of religion in the nuclear decision-making of Pakistan requires an understanding of how religion is significant in the society and state policies of Pakistan in general. Pakistan is a Muslim majority state that was created based on the idea of the two-nation theory proposed by Muslim League in the last decades of independence struggle against Britain. The two-nation theory proposes that identities of Muslim and Hindu constitute two different nationalities, and they cannot live within the nation-state. Muslim League considered the religious identity in South Asia as equal to ethnic and linguistic identity in Europe such as French and German. According to this logic, just like two different ethnic or religious

86

identities constitute two distinct nations in Europe two religious identities constitute two nations in South Asia.

The opposition to this theory comes from two aspects. First, two separate nations are not a necessary outcome of difference in religious identity. Various religious people can live together in a single nation-state. It is the basic idea of India. Second, the religions of South Asia, Hinduism or Islam, are not homogenous. There are different ethnic and linguistic groups within each religion. The creation of Bangladesh after 1971-War strengthened this criticism. Therefore, security threats to the core idea of two-nation theory may be through the attempt of India to integrate Pakistan with it or to dismember Pakistan into different nations. The underlying security of Pakistan is to preserve its Muslim identity different from India and use that identity as a binding force among various ethnic and linguistic groups without disintegration of the state.

Even though Pakistan was created based on Muslim identity, the idea of a nation-state based on any identity was rooted in Western values, not in Islamic principles. Because, Islam does not promote the creation of separate nation-states according to religious identities. The demand for Pakistan itself was opposed by a majority of Islamic scholars during the time of Pakistan movement (Abbas 2008: 53). For Muslim League and Muhammed Ali Jinnah, Pakistan was a state for Muslims rather than an Islamic state (Cohen 2004: 161). The formation of Pakistan was based on 'Muslim' as an identity but not based on 'Islam' as a code of conduct. Muslim League and Jinnah wanted that Pakistan is to be ruled by Muslims but not necessarily based on laws of Shari'a. Even though meanwhile Pakistan was ruled by leaders with the agenda of Political Islam, the role of religious principles in the decision-making of Pakistan is still marginal. For example, the decision of Zia-ul-Haq, who is known as the most Islamist rulers of Pakistan, to exercise capital punishment against Z. A. Bhutto was not based on Islamic principles, but the decision was against Shari'a (Kaushik and Mehrotra 1980: 125).

Pakistan was not ruled by Islamic jurisprudence, but "Islam was used and being used, as an instrument to maintain the privilege of the privileged" even by leaders who are secular in their life (Ayub 1979: 545). It distinguishes Pakistan from Iran which is ruled by the Supreme Leader who is *Velayat-e-Faqih* (guardian jurist). The Guardian Council with 12 jurists has ultimate authority in Iran. In Pakistan, the decisions are taken by either civilian or military leaders, not by any Islamic jurists. In short, Pakistan is still not an

"Islamic state", but it is only a "Muslim state" whose citizens are entirely or predominately Muslim" (Cohen 2004: 162).

Role of Religion in Nuclear Decision-Making

Religion has different aspects. It is a code of conduct; it is the identity of people, it is part of civilization and culture. Regarding the role of the Islamic code of conduct in nuclear decision-making, it has been established in this book's second chapter "Role of Religion in Security Policies of the States" that neither Sunni nor Shia scholars have promoted the development of nuclear bomb as a duty of state leaders. They have a different opinion on whether the nuclear weapons are allowed or not. The view of prohibiting nuclear weapons has not constrained Pakistan. It is mainly because of lack of powerful constitutional role to Islamic jurisprudents in the policymaking of Pakistan. In addition to that Pakistan can also follow the opinion of scholars who justified nuclear weapons if it is possessed by the adversary.

However, these scholars also allowed only possession of nuclear weapons for deterrent purpose, and few of them allowed the use of it for second strike purpose only. If the policies of Pakistan are constrained by religious principles, it has to follow a no-first-use policy which Pakistan has not yet explicitly declared. The lack of powerful constitutional role to Islamic jurisprudence in policymaking keeps the possibility of first use of nuclear weapons alive. In short, the development of nuclear weapons in Pakistan is not motivated nor constrained by religious principles. Since the constitutional role of religious jurisprudents is very weak, it is not sure whether the religion will constrain the use of the nuclear weapons.

Although religious principles do not promote nuclear weaponization, the religious identity has an influential role in nuclear decision-making. The preservation of this identity and prestige over it have a decisive role in the nuclear policies of Pakistan. It also has a role in constructing friends and enemies and in shaping public opinion. This influence can be analyzed by using various approaches of nuclear proliferation.

According to realism, which is the dominant explanation of Pakistan's nuclear policies, the weaponization was to balance against security threat from India. Since national security is defined in realist view as to protect its internal values from external threat, the security of Pakistan is to protect its core idea of the nation, i.e., two-nation theory and its Muslim identity distinguishing from India. Pakistan developed nuclear weapons to avoid repetition of the experience of 1971 war and to ensure the survival of state with Muslim identity without further dismemberment and without integrating to India. Since religion is the base for the creation of Pakistan, its perception of the security threat from India was also influenced by religion. The conflict between these two countries was presented in Pakistan as a war between "Pakistan Muslim Us" versus "Hindu Indian Them" (W. S. Khan 2010: 3). So, Pakistani leaders regarded Indian threat not just in material terms such as the territorial dispute over Kashmir, but also in ideological terms of religion. For Pakistani elites, this portraying India as a threat to Islam and Muslim-Pakistan was helpful to integrate the nation which does not have any common identity other than religion.

Stephen Walt's explanation of "balancing against threat" is useful to understand the influence of religion through the prism of realism. Unlike the Waltzian concept of "balance of power", S. Walt argues that states are balancing against the threat. In the case of South Asia, Pakistan is balancing against India is not just because of India is a powerful state, but also because it is a threat to Pakistan. The religious backgrounds of these countries have a decisive role in formulating this threat. Even though India is a secular multi-religious country, Pakistan considers the conflict between two nations as "Hindu-India" versus "Muslim-Pakistan." The formation of Pakistan in the name of a religious identity and its view about India as a threat to nation-building based on that identity constitutes the basis of conflicts between these two neighbouring states.

Copenhagen School of Constructivism informs the significance of religion in securitizing India as a threat to state and society of Pakistan. Since its establishment, Pakistani leaders believe that India has determined "to destroy the separate political and cultural identity of the Muslim community in the subcontinent" (Hilali 2002: 3). They cite 1971-War with India as an attempt by India to destroy Muslim unity and identity of Pakistan. Therefore, nuclear weapons are introduced as an effective deterrent against India to defend state and society of Pakistan.

Constructivism is also helpful to identify how religion, among many other factors, plays a role in constructing the identity and interests of India and Pakistan. The conflicting identities of India and Pakistan create conflicting interests. These identities and interests are shaped by historical experiences such as the 1971-War. At the same time, the shared identity of religion has a role in getting financial support from some Muslim countries to Pakistan. Religion also plays a significant role in constructing the idea of Pakistan as a Muslim-state. The nationalism, which is defined as a feeling of belongingness to an "imagined community", is also based on religion in Pakistan. Pakistani nationalism was constructed with its Muslim identity and Pakistani rulers considered themselves as successors of Muslim rulers of the subcontinent (F. H. Khan 2005: 4). Pakistani elites have connected supporting nuclear program as equal to patriotism. If anybody opposes the nuclear plant, he will be treated as traitor of Pakistan (Hoodbhoy and Mian 2014: 1134).

However, there is a difference between Islamic nationalism and Islamic internationalism (*ummatism*) (Dajani 2011). While the Islamic internationalists aim to establish one global Muslim society without national boundaries, Islamic nationalists are "ethnocentric religious organizations that have an ethnically oriented religious agenda" (Dajani 2011: 4). Jamaath-e-Islami and al-Qaeda are examples for the proponents of Islamic internationalism or *ummatism*. Pakistan was created based on Islamic nationalism. There was a clash between Islamic nationalist ideology of the Muslim League led by Muhammed Ali Jinnah and Islamic internationalist ideology of the Jamaath-e-Islami when demand for a separate nation of Pakistan was made. This difference between Jamaath-e-Islami and Government of Pakistan still continues. This difference leads to a confrontation between the Jamaath leaders and government from time to time (Dajani 2011: 11). The understanding of this difference between Islamic nationalism and *ummatism* is necessary to avoid the confusion over whether the Pakistani bomb is for defending all Muslim states. The statement of Jamaath leaders is based on their ideology of *ummatism*, and that is different from Pakistan government's policy. Not only secular leaders, but also Islamic nationalist leaders of Pakistan do not share the view of Jamaath-e-Islami or other Islamic internationalists to use Pakistan's nuclear weapons for extended deterrence of all Muslim states.

At the domestic level, religion has a role in forming a strategic culture of Pakistan since Islam "contributes to shaping societal dispositions and the orientations the policymakers" (Rizvi 2002: 319). Feroz H. Khan (2005: 4) has identified the role of religion in formulating an image of Pakistan as a nation. Peter R. Lavoy (2006: 14) has also indicated the role of Muslim nationalism as a key element of Pakistan's strategic culture. Pakistani civilian and military leaders have used Islamic symbols as a source of inspiration. However, as argued before, the Islamic principles were interpreted and used by military and civilian leaders as per their preferences, and Islamic scholars have no constitutional role in defining these principles.

At the individual level, mythmakers have used religion to popularize their position. The statement of Bhutto calling Pakistani bomb as an Islamic bomb is an instance. Instrumentalist approach can explain how political leaders use religion for achieving other goals. Mohammed Ayub (1979) argues that Pakistani leaders from Muhammed Ali Jinnah were using religion for other purposes such as providing legitimacy to government policies and for justifying political injustice and unjust rule. Even the leaders who follow secular views in personal life used religion in the public sphere for other purposes. For example, by portraying Pakistan's bomb as "Islamic Bomb," Bhutto tried to achieve his personal interests. One such purpose was to get the support of Islamic fundamentalists who were opposing Bhutto for his policies. Writing while waiting for his death in prison, the term "Islamic Bomb" might help Bhutto to portray his opponents as against the interest of Islam and to get support from Islamic states such as Saudi Arabia to put pressure on Zia-ul-Haq to release him. The use of the term "Islamic Bomb" also serves national interests in many ways. It will be discussed in detail in the following section.

Is Pakistan's Bomb an "Islamic Bomb"?

The phrase "Islamic bomb" first appeared in the early 1970s in a private conversation of Z. A Bhutto (Sciolino 1998). According to Richard Bonney (2012: 1), the concept was referred by Bhutto in 1965. However, the term emerged as a matter of concern and an interesting theme in the academic circles much later, only when Bhutto wrote from the prison that

> "We know that Israel and South Africa have full nuclear capability. The Christian, Jewish and Hindu civilizations have this capability. The communist powers also possess it. Only the Islamic civilization was without it, but that position was about to change" (Bhutto 1979: 151).

The term became titles of books co-authored by D. K. Palit and P. K. S. Namboodiri (1979) and by Herbert Krosney and Steven Weisman (1981). BBC produced a documentary entitled "Islamic Bomb" in 1979 (Hoodhoy 1993: 42). Sohail Hashmi (2004: 337) points out two ironies in the origin of the term. First, secular Bhutto would be known for coining the term "Islamic bomb." Second, the concept of Islamic bomb served the interest of Zia's regime that executed Bhutto. The concept was later used by many political leaders and newspapers. Iran's foreign minister Kamal Kharrazi expressed support for Islamabad in after 1998-test in the name of Islamic solidarity. Saudi Arabia supplied with 50,000 barrels per day of free oil for overcoming international sanctions. An enthusiastic mob led by Jamaath-e-Islami celebrated in the street claiming the Pakistani bomb as a bomb for entire Muslim *Ummat* (Hoodbhoy and Mian 2014: 1135).

Even though the term was introduced by Z. A. Bhutto, later Pakistani leaders like Nawaz Sharif and Pervez Musharraf denied the pan-Islamic aspect of Pakistani bomb and argued that portraying it as an "Islamic bomb" is Zionist and Indian propaganda for rallying the Western states against Pakistan (Hashmi 2004:341). Gawhar Ayyub Khan contended that "we must not call it 'Islamic bomb', for there is nothing called a Christian bomb, Jewish bomb or a communist bomb" (Hashmi 2004:341). When Iranian foreign minister described Pakistan's bomb in terms of Islamic identity, Nawaz Sharif, then Prime Minister, rejected such an image and indicated that "bombs do not have religious identities, only national identities" (Yasmeen 2001: 208).

These arguments and counter-arguments make it imperative a deeper analysis of factors that contributed to the emergence of the term "Islamic bomb" and counter-narratives to these factors. First is the financial contribution of Muslim states to Pakistan. This financial support is a major argument of D. K. Palit and P. K. S. Namboodiri (1979) to establish Pakistan's nuclear weapons as Islamic bomb. Yossef Bodansky (1998) also argues that Pakistan enjoyed active support from Arab states, particularly Libya and Saudi Arabia (Bodansky 1998: 1). However, this financial assistance does not provide an indication about the religious aspect of Pakistan's nuclear program. Because even though Pakistan benefited from the financial support of Arab countries, it got more important aspects of nuclear weapons from other places: China supplied blueprints, enriched uranium, and scientific know-how; the technology was stolen from the Netherlands; the US gave consent and financial and military support during the time of Afghanistan war (Sciolino 1998:2).

The second argument is the usage of the term by Pakistan's leaders such as Z. A. Bhutto. Brij M. Kaushik and O. N. Mehrotra (1980: 90) figure out four motivations for Pakistani leaders to link its nuclear bomb with Islamic civilization. First is to get the leadership of Islamic bloc. Second is to get financial support oil-rich Muslim countries. Third, it may be a tactic of Pakistani leadership to divert pressure and criticism of anti-proliferation regime towards all Islamic countries. Fourth, Pakistani leadership has sought to whip up anti-Zionist aspect of the bomb. Palit and Namboodiri (1979: 9-10) also point out similar motivations. However, all these purposes, may be with the exception of anti-Zionist agenda, are clearly related Pakistan's national interests such as leadership, prestige, economic benefits, and reduction of international pressure. Apart from these national interests, Z. A. Bhutto had individual motivations such as getting the support of Islamic fundamentalists at domestic level and the Muslim states at international level. In short, the motivation of Bhutto or others by using the term "Islamic bomb" was not to promote a pan-Islamic agenda but to achieve national and individual interests.

The third concern about the existence of "Islamic bomb" is transferring nuclear know-how to Islamic states, especially to Iran and Saudi Arabia, as a return for their financial assistance. Yossef Bodansky argues that "Pakistan's contributions to the nuclear programs of the Islamic Republic of Iran date back to the early 1980s" (Bodansky 1998: 2). However, in the following line, when Bodansky says Iran got support from Pakistan and France, he is biased in his argument that the motivation of Pakistan was religious purpose and that of France was material interests. Another weakness of Bodansky's argument is his allegation that Pakistan had supported both Iran and Iraq in their nuclear program in the 1980s. It should be noted here that Iraq and Iran were fighting a bloody war in the 1980s and most of the Arab Muslim states including Saudi Arabia and western countries including the US were in support of Iraq. Since Pakistan was the beneficiary of both Saudi Arabia and the US, supporting Iran would have affected it negatively. Furthermore, even if Pakistan supported Iran or Iraq, it would not serve Islamic interest since two Muslim countries were fighting against each other, and helping them cannot be termed as Islamic. So, to deduce an argument for the "Islamic bomb" based on Pakistan's support for Iran or Iraq in the 1980s do not hold logical ground and must be considered a fallacy.

It should also be noted that even though Pakistan and Iran had a good relationship during the Shah's period since both countries were members of Western alliance such as SEATO and CENTO, the Islamic revolution in Iran led to a clash of interests (Hoodbhoy 2013: 155-156). Both countries competed for getting influence in Afghanistan. Pakistan's support to Taliban, which was anti-Shia in its ideology, affected its relation with Iran negatively. Even Zia-ul-Haq, who was known for his Islamist ideology, rejected Iran's request for access to Pakistan's nuclear technology. It was an indication of Pakistan's consideration of its nuclear weapon as "instrument of national power, not as a means of altering the international balance in favor of Islamic states" (Yasmeen 2001: 203-204). Aslam Beg, who was the Chief of Army Staff, had also denied Pakistan's support to the nuclear program of Iran and called such allegation as Jewish conspiracy (Abbas 2008: 256).

Even at present, Pakistan has to be concerned about the reaction of Saudi Arabia, which is its main financial supporter, apart from American response. The clash between Sunni and Shia branches of Islam is also relevant in Pakistan's nuclear policy towards Iran. Hassan Abbas (2008: 147) quotes IISS report saying that,

> "Pakistani officials across the board insist that Zia did not approve any nuclear dealings with Iran that would involve the provision of sensitive technology. They argue that his strong Sunni beliefs and his strategy to increase the role of Sunni Islam throughout Pakistani society and official institutions put him at odds with Iran's Supreme Leader Ayatollah Ruhollah Khomeini and made any sensitive dealings with Iran very unlikely".

Pakistan is also worried about losing its status as the sole Islamic country with nuclear weapons. Pervez Hoodbhoy quotes Wikileaks report on Pakistan's efforts under General Pervez Musharraf to dissuade Iran from pursuing nuclear weapons (Hoodbhoy 2013:158). In 2006, Khurshid Kasuri, who was Pakistan's Foreign Minister then, said that "we are the only Muslim country [with such weapons]," he said, 'and don't want anyone else to get it" (Hoodbhoy 2013:158).

The transfer of nuclear technology to Saudi Arabia is another concern over Pakistan's weapons. The financial support of Saudi Arabia to Pakistan is a major reason for this concern. Pakistan received more aid from Saudi Arabia than from any other country (Hoodbhoy and Mian 2014: 1136). Since it is hard for Saudi Arabia to get nuclear technology from other sources, media and policymakers suspect that it may turn to Pakistan for fulfilling its nuclear ambition. However, considering Pakistan's desire to continue as the sole Muslim nuclear state and due to other strategic reasons, it is unlikely to transfer its nuclear technology to Saudi Arabia. Mark Fitzpatrick (2015) contends that "Pakistan has strong strategic, political and economic incentives to keep its nuclear weapons to itself" (Fitzpatrick 2015: 107). It should also be noted that Pakistan has refused the request of Saudi Arabia to send its military to Yemen for the Saudi intervention there. It is sure that nuclear technology transfer to Saudi Arabia will provoke its neighboring country Iran and larger international community and it will bring new troubles for Pakistan (Hoodbhoy and Mian 2014: 1136). Fitzpatrick argues that since Pakistan has a powerful adversary, India, on its eastern side, it would be a strategic blunder to involve in Saudi-Iran dispute (Fitzpatrick 2015: 107).

Fourth and most repeated concern over Pakistan's nuclear bomb is that it might be used to protect all Muslim countries by deterring outside attack including Israel and the US. Arguing anti-US aspect of the bomb, Bodansky (1998: 4) says that "Islamabad is gradually shouldering its responsibilities as the sole declared Muslim nuclear power fronting for the entire hub of Islam against the US-led West and its nuclear might as well as standing up for the "honor" of Islam." He also argues that Islamabad is willing to provide extended deterrence balance against Israel and the US. Herbert Krosney and Steven Weisman (1981) also present Pakistan's bomb as a significant deterrent in Arab-Israel conflict. These arguments presume pan-Islamic motivation in Pakistan's nuclear program. However, as Pervez Hoodbhoy argued, "nothing in the history of Pakistan has shown a substantial commitment to any pan-Islamic cause" (Hoodbhoy 2010: 78).

Regarding implication of Pakistan's nuclear weapons in the Arab-Israel conflict, Akhtar Ali contends that the possession of a nuclear weapon by a Muslim country may create uncertainty. However, the physical transfer of nuclear weapons to the Arab may never take place. Such a transfer next to impossible especially considering the reaction of the US and Israel to such as transfer (Ali, 1984:95). A direct attack from Pakistan to Israel can be possible if it develops substantial ballistic missiles. Israel came under missile range of Pakistan only when it tested Shaheen III in March 2015. However, the accuracy of the missile to attack Israel is doubtful considering the geographical location and size of Israel which is a small country surrounded by Muslim-majority states. Israel has also developed its Ballistic Missile Defence (BMD) technology.

The argument for Pakistan's ability to deter the US is very weak. Pakistan joined hands with the US during and after the Cold War. This alliance with the US includes support for the US invasion of Muslim majority countries. Even after the development of nuclear weapons, Pakistan was not capable of deterring the threat of US when it invaded Afghanistan. Pakistan was compelled to make a U-turn from supporting the Taliban regime to support the US in its 'Global War on Terror'. It shows that Pakistan's nuclear weapons are not capable of standing against the US and deterring it. Pakistan also sided with the US when it invaded Iraq. So, Pakistan is very unlikely to risk its national security by causing retaliation from the US by transferring nuclear technology to West Asian states.

Conclusion

The nuclear policies and doctrines of Pakistan are Indo-centric. The nuclear weaponization was mainly motivated by security threat from India. In the first two decades of the independence, Pakistan tried to balance against India through modernizing conventional weapons and external balancing by allying with the Western states. However, the defeats in the wars against India and lack of enough support from Western countries led Pakistan to change strategy to internal balancing by developing nuclear weapons. Apart from security motivation, different theoretical frameworks indicate the role of some other factors as well. These factors include international status, parochial interests of bureaucrats and military, domestic strategic culture, and individual oppositional nationalists and mythmakers.

This case study focused on the role of religion in the nuclear weaponization of Pakistan. It concludes that the primary cause of nuclear weaponization was not religion, instead it was security threat from India. Pakistan would have developed nuclear weapons even if it was a Christian or secular country. However, religion played a role as an intervening variable by constructing the hostile relationship between India and Pakistan. The national security of Pakistan was to secure the basic value of the nation, i.e., the two-nation theory. Therefore, any security threat from India was interpreted through the framework of this theory and Muslim identity of Pakistan. Realism, especially Stephen Walt's concept of balance against the threat, and constructivism can explain this role of religion. The strategic culture of Pakistan was influenced by religion. Religion also played a role in creating nuclear nationalism in the country. Individual leaders also used religion for nuclear mythmaking and for mobilizing people to support the nuclear weapon program. All these influences of the religion were in terms "Muslim" as an identity, not "Islam" as a code of conduct. Unlike Iran, religious jurisprudents have no constitutional role in determining whether one act is valid in Islamic Shari'a or not. Therefore, religious scholars' opinions on prohibiting and use of nuclear weapons are not constraints in nuclear policies of Pakistan. Although religious identity has a role in nuclear weaponization as an intervening variable, it cannot be termed as an "Islamic bomb" and it is only for the national purpose of Pakistan. Pakistan is unlikely to transfer nuclear technology to the other Muslim states or to provide them an extended deterrence due to possible negative impacts of such an action on its national security. Even if it transfers to any state, it will not be motivated by religion but by national or individual interests which are nonreligious.

Conclusion

The aim of this study was to explore the role of religion in the nuclear decision-making of states. It explored the influence of religion at various levels such as structural, domestic, and the individual. This study also attempted to learn whether religion has an influence on nuclear policies, and, if yes, why does the same religion motivate different states to act in opposite directions. After rigorous case studies, this study concludes that the nuclear decision-making of Iran is constrained by religious principles which have a constitutionally powerful role. In the case of Pakistan, even though the security threat from India is the primary cause in its nuclear weaponization, religious identity plays the role of intervening variable in different ways.

Nuclear policies of all states cannot be explained based on a single variable or by a single theory. The factors behind the policy are different from one state to another. The nuclear decision-making of a single state could also be shaped by multiple factors. The first chapter of this book explained in detail the factors influencing nuclear policies through the prisms of different theories like structural, domestic, and individual analytical levels. This book explained the role of religion in shaping domestic norms and policies of states with a case study of Iran. Accommodating religion into the paradigms of International Relations and Security Studies can broaden the frameworks of existing structural, domestic, and individual level analyses, for religion is a significant factor in shaping the psychology of an individual, public opinion and strategic culture, and in constructing friends and foes.

Religion has been an ignored factor in the literature of Social Sciences including International Relations due to many factors which were discussed in detail in the chapter on "Role of Religion in Security Policies of the states." They include the social situation during the origin of Social Sciences, focus on Western European society and influence of behavioralism in Social Sciences. However, religion has re-emerged as a significant factor in the last few decades. Policymakers and scholars realized the importance of religion in shaping the behavior of people. The resurgence of religion and revival of religious movements in many parts of the world, the thesis of "clash of civilization" by Samuel P. Huntington, and the 9/11 attack on the USA were some of factors that made religion more visible in the discourse of Social Sciences. Western scholars realized the limitations of using only the Euro-centric prism to understand politics and culture of African and Asian societies. It led them to consider Eastern civilizations with respect to their distinctive characteristics, including the public role of religion.

Religion plays an influential role in international relations. It shapes personalities and preferences of policymakers; it is a base for identity in nation-building; it legitimizes/criminalizes the actions of leaders at domestic and international levels; it contributes to international norms such as code of conduct for war and human rights. However, compared to other disciplines such as political science and sociology, the discipline of IR was very late to accommodate religion into its theoretical frameworks. Even though religion continued as a marginalized variable at the practical level, it can be accommodated within the framework of existing theories. Among various branches of realism, classical realism can explain the role of religion in shaping the nature and interests of individuals. Stephan Walt's explanation of structural realism, which says that states balance against the threat, not against power, is useful to identify the role of religion in constructing threat. Constructivism can also accommodate religion as a variable to construct identities of other states as friends or enemies. The feeling of belongingness to a particular religious identity can be a base for the development of nationalism and nation-state. Religion also contributes to the origin of various ideas and norms at domestic and international levels. The framework of liberalism is also amendable to accommodate religion since neoliberalism recognizes multiple actors and agendas in international politics. It can explain the role of transnational religious organizations at international level and religious pressure groups at domestic level. Religion also contributes to the development of international regimes. The idea of "soft power" is another concept in neoliberalism to identify the influence of religion which is a central pillar of civilizations.

This book analyzed the role of religion in nuclear policies of Iran and Pakistan by using these broadened theoretical frameworks of IR. Since Iran and Pakistan are Muslim majority countries, understanding Islamic view on nuclear weapons is inevitable. Islam is one of the most influential religions in the world. Since Islam does not separate public and private spheres of life, many Muslim states adopt its principles for policymaking including security and foreign policies.

Islam puts forward many ethics related to war and use of force. Based on these ethics, its views on nuclear weapons can be analyzed. There are two opinions on the legal aspect of nuclear weapons among the learned *Ulama* (Islamic scholars) of both Shia and Sunni sects. The first position prohibits both possession and use of nuclear weapons (*Haraam*) and the second position allows the possession, not the use, of nuclear weapons for the purpose of deterrence (*Mubah*). So, Islam, as a code of conduct, either motivates non-weaponization or just allows development of weapons for deterrence. Even though there is no consensus among religious scholars regarding the lawfulness of nuclear weapons, the recognized Islamic scholars do not consider the development of nuclear weapons as state leaders' duty, neither *Vaajib* (the obligatory) nor *musthahabb* (the recommended). Their disagreement is only on whether or not the possession of nuclear weapons is allowed. Most of the Sunni and Shia scholars unconditionally prohibit the possession of nuclear weapons. The scholars, who allow the development of nuclear weapons, have different opinions regarding their use and some of them permit their use only for a second-strike purpose. It is near to consensus that there is a prohibition on the use of nuclear weapons for the first strike.

The Supreme Leaders of Iran, both Ayatollah Khomeini and Ayatollah Khamenei, and most of the Ayatollahs opine that the possession and use of nuclear weapons are prohibited by Islam. Their fatwas regarding nuclear weapons have a significant role in shaping the nuclear policies of Iran and its non-weaponization. Iran's nuclear program was introduced during the time of the Shah. The declared motivation of the Shah was to use nuclear energy for national economic development. He rejected the plan for nuclear weaponization considering it would negatively affect the security of Iran and its relationships with the western states. Religion was not an influential force in Shah's nuclear program or non-weaponization. The Islamic revolution of 1979 made dramatic changes in foreign and security policies of Iran including nuclear program. The Islamic regime of Ayatollah Khomeini, and later Ayatollah Khamenei, opposed Weapons of Mass Destruction on ideological base. Based on various evidence referred in the chapter on Iran, this study concludes that Iran has not yet developed a nuclear weapon, despite its technological capability.

It is difficult to explain the absence of weaponization in Iran without considering the role of religion in nuclear decision-making, for the explanation of Iran's nuclear policies based on other factors, such as security threat, balance of power, trust in international regimes and norms, technological capability, and nationalist feeling of leaders, leads scholars into predicting nuclear weaponization. The theories of nuclear proliferation would continue to be flawed, as in the case of Iran, unless the role of religion is considered seriously. As a theocratic state, the religious principles and interpretations of the Supreme Leader and the Guardian Council, which consists of 12 Islamic jurisprudents, have a constitutionally powerful role in Iran. The position of the Supreme Leader as 'guardian jurist' makes his fatwa politically and religiously final word in Iran. His fatwa binds all departments of the state and overrides all other military and political considerations. The Guardian Council has veto power over the executive and legislative decisions. Since the Supreme Leader and Guardian Council consider nuclear weapons as against Islamic principles, any law to develop them would be *null* and *void* according to Iranian constitution. So, the religious judgment of the Supreme Leader and Guardian Council is more powerful than a mere legislation. In addition, a fatwa of Ayatollah Khamenei against nuclear weapons cannot be equalled with a fatwa of any other qualified Islamic scholar which binds only to his followers. Similarly, as the Supreme Leader is Commander in Chief of the military, his opinion overrides the interests of other military leaders.

Even if a radical political leader comes to power, he cannot change Iran's policy on nuclear weapons unless the religious position of the Supreme Leader and the Guardian Council changes, something more difficult than an amendment in secular law. Since the legitimacy of the government among people is rooted in religion and norms against nuclear weapons popularized through repeated religious statements, it is hard for the Iranian state to reverse its position. So, leaders would not bluff by making repeated religious statements against nuclear weapons if Iran really wants them. Therefore, a proper understanding of Iran's nuclear policy requires consideration of the influence of religion in shaping domestic norms, worldview and behaviors of individuals.

The role of Islam in Pakistan, even though it is a Muslim state, is different from that of Iran. There is no religious authority in Pakistan with the power of the Supreme Leader or Guardian Council of Iran. The policies of Pakistan are decided either by civilian or military leaders. Therefore, these leaders interpret Islam according to their interests. In other words, Islam does not rule Pakistan, instead Pakistani leaders, including those who are secular in their approach, use religion for other purposes. So, the influence of religion in the nuclear decision-making of Pakistan is different from that of Iran.

The nuclear policy of Pakistan, just like its other security policies, is Indo-centric. The orientation of nuclear program in Pakistan shifted from civilian to military purpose in the 1970s due to many factors such as the defeat to India in the war of 1971, the election of Zulfikar Ali Bhutto to the post of President and the 'Peaceful Nuclear Explosion' of India in 1974. Due to the central role of security threat from India, not just realism, but other theories have also taken the security concern as an important variable in the nuclear weaponization of Pakistan. Out of the different theories of its weaponization, security-oriented explanation of realism is the dominant one. According to realism, the internal balancing through nuclear weapons was the only option for Pakistan to counter India after experiencing defeats in three wars and failure of early strategies such as external balancing through alliance building and internal balancing through conventional weapons. In addition to the security consideration, other factors such as prestige, parochial interests of bureaucrats and politicians, strategic culture, and mythmaking and nationalist feeling of individuals are also key variables which are used to explain the nuclear weaponization of Pakistan.

This study concludes that it was the security threat from India, and not religion, which became the primary reason behind Pakistan's nuclear weapons. It would have developed a nuclear weapon even if it was a Christian or a secular state. However, it does not mean that religion had no role to play in the nuclear development of Pakistan. Although Islamic principles do not have impacts on nuclear policies, the 'Muslim identity' is a significant factor in nuclear decision-making. Religious identity is the base for the formation of the state and nationalism in Pakistan. The basic principle behind the formation of Pakistan in 1947 was the two-nation theory which proposed separate nation-states based on religious identities. After that, the clash between India and Pakistan was narrated in terms of religious identities such as Hindu India versus Muslim Pakistan. The meaning of national security in Pakistan is to secure this basic value of the nation, i.e., the two-nation theory. So, the security threat from India, which was the dominant factor in the nuclear weaponization of Pakistan, was also influenced by religious identities of these countries. In addition to the formation of identity and the construction of security threat, religion influences the shaping of strategic culture of Pakistan. Religion is also used by civilian leaders for nuclear mythmaking and mobilizing public support in favor of weaponization. So, religion is an intervening variable in the nuclear policies of Pakistan.

All these influences of religion in Pakistan's policies are in terms of 'Muslim' as an identity, and not 'Islam' as a code of conduct. Due to this, as explained earlier, nuclear weapons are not promoted by Islamic jurisprudence according to any of the two qualified interpretations. As per one of these interpretations, development and use of nuclear weapons are prohibited unconditionally. It is clear that Pakistan's nuclear policies have not been constrained by this opinion. The scholars, who have allowed the development of nuclear weapons only for the deterrence purpose, also prohibit their use for the first strike. If Pakistan follows the second opinion, it has to adopt a "no first use" policy, but it has not declared such a policy. It shows that, unlike Iran, religious ethics of warfare do not constrain nuclear decision-making of Pakistan. This is mainly because of the lack of a constitutionally powerful role for the Islamic jurisprudents to determine whether or not an act of the government is valid as per Islamic Shari'a. The decisions in Pakistan are taken by either civilian or military leaders. There is no religious authority in Pakistan with the same power of the Supreme Leader or Guardian Council of Iran. Pakistan is only a 'Muslim majority' state, not an "Islamic state" in contrast to Iran. Islamic Shari'a does not rule Pakistan, but Islam is used by the political and military leaders, including those who are secular in their personal life, for their individual and national interests.

The term "Islamic bomb", coined by Z. A. Bhutto, is an example of using religion for individual interests such as getting financial support from Arab countries, leadership role among Islamic states, and Bhutto's agenda to get the support of Muslim states at the international level and the Islamic groups at the domestic level. The term "Islamic bomb" indicates that the nuclear weapons of Pakistan are to defend all Muslim states and *Ummah* either through extending deterrence or transferring technology. However, Pakistan is unlikely to transfer nuclear technology to other Muslim or non-Muslim states or to provide them an extended deterrence if it affects the security of Pakistan adversely. It has been discussed in detail in the chapter "Role of Religion in Nuclear Policies of Pakistan" that transfer of nuclear technology to other countries including Iran and Saudi Arabia will have negative impacts on the national security of Pakistan. So, this study concludes that the nuclear bomb of Pakistan serves only that country and not the whole Muslim community. Hence, it cannot be called as an "Islamic bomb."

In short, this study concludes that religious norms have a decisive role in the nuclear decision-making of Iran. Iran would have gone for weaponization if it has not been constrained by religion. At the same time, since it does not have a constitutionally powerful religious authority, Pakistan is not constrained by Islamic views on nuclear weapons. The most significant factor in the nuclear decision-making of Pakistan is security threat from India. However, religious identity works as an intervening variable in different ways. The nuclear weapon of Pakistan is only meant to serve its national interest, and it is not an "Islamic bomb". It is unlikely for the former to provide extended deterrence or to transfer nuclear technology to other Muslim countries due to its harmful impacts on Pakistan's national interests.

This study assesses different ways to understand the influence of religion in the decision-making of modern states. In doing so, it also tried to broaden the existing theoretical frameworks on religion, especially, in a context in which the latter has returned to the international relations in incredibly forceful ways. Although this study deals with the role of religion in nuclear decision-making with case studies of only Iran and Pakistan, it could be extended to other policies and to other states.

References

(* Denotes Primary Sources)

Abbas, H. (2008), *Causes That Led to Nuclear Proliferation from Pakistan to Iran, Libya, and North Korea: Investigating the Role of the Dr. Abdul Qadeer Khan Network*, Ph.D. Thesis, The Fletcher School of Law and Diplomacy, Tufts University.

Abbasi, R. K. (2010), *Understanding Pakistan's Nuclear Behaviour (1950s–2010): Assessing the State Motivation and its International Ramifications (a Three Models Approach)*, Ph.D. Thesis, Leicester: University of Leicester.

Abbasi, R. K. (2012), *Pakistan and the New Nuclear Taboo: Regional Deterrence and the International Arms Control Regime*, Bern: Peter Lang.

Abraham, I. (2009), "Introduction: Nuclear Power and Atomic Publics" in Itty Abraham, *South Asian Cultures of the Bomb: Atomic Publics and the State in India and Pakistan*, Bloomington & Indianapolis: Indiana University Press.

Ahmed, S. (1999), "Pakistan's Nuclear Program: Turning Points and Nuclear Choices", *International Security*, 23(4): 178-204.

Ali, A. (1984), *Pakistan's Nuclear Dilemma: Energy and Security Dimensions*, New Delhi: ABC Publishing House.

Alidoost, A. A (2014), "The Religious Foundations of the Edicts (fatawa) by Shi'ite Jurists Prohibiting Weapons of Mass Destruction", Summary of paper presented at Conference on "Nuclear Jurisprudence" in March 2014, Tehran.

Allison, G. T. (1971), *Essence of Decision: Explaining the Cuban Missile Crisis*, Boston: Little, Brown.

Altman, A (2009), "Ayatullah Ali Khamenei: Iran's Supreme Leader", *TIME*, 17 June 2009.

Anderson, B. (1983), *Imagined Community: Reflections on Origin and Spread of Nationalism*, London: Verso

Ansari, A. M, (2013), "To be or not to be: Fact and Fiction in the Nuclear Fatwa Debate", [Online: web] Accessed on 30 March 2016, URL: https://rusi.org/commentary/be-or-not-be-fact-and-fiction-nuclear-fatwa-debate.

Asad, T, (1993), *Genealogy of Religion; Discipline and Reasons of Power in Christianity and Islam*, Baltimore and London: The Johns Hopkins University Press

Ayoob, M. (1979), "Two Faces of Political Islam: Iran and Pakistan Compared", *Asian Survey*, 19 (6): 535-546.

Bahari, M. (2008), ''The shah's plan was to build bombs", An Interview with Akbar Etemad, *New Stateman*, 11 September 2008, URL: https://www.newstatesman.com/asia/2008/09/iran-nuclear-shah-west.

Bahgat, G. (2013), "Dealing With Iran: The Iranian Nuclear Crisis: An Assessment", *Parameters* 43(2): 67-76

Barnett, M (2011), "Social Constructivism" in John Baylis and Steve Smith (eds.), *The Globalization of World Politics: an Introduction to International Relations*, Oxford: Oxford University Press.

Behestani, M. and Shahidani, M.H. (2015), "Twin Pillars Policy: Engagement of US-Iran Foreign Affairs during the Last Two Decades of Pahlavi Dynasty", *Asian Social Science*, 11 (2): 20-31.

Berger, T. U. (1996), "Norms, Identity, and National Security in Germany and Japan", in Peter J. Katzenstein (ed.) *The Culture of National Security: Norms and Identity in World Politics*, New York: Columbia University Press

Bernstein, J. (2014), *Nuclear Iran*, Cambridge, Massachusetts, and London, England: Harvard University press.

*Bhutto, Z. A. (1967), *The Myth of Independence*, Reproduced by: Sani H. Panhwar, www.bhutto.org, [Online: web] Accessed on 27 April 2016, URL: http://bhutto.org/Acrobat/Myth%20of%20Independence.pdf

*Bhutto, Z. A. (1979), *If I Am Assassinated*, Reproduced in PDF Format By: Sani Hussain Panhwar, www.bhutto.org, [Online: web] Accessed on 27 April 2016, URL: http://bhutto.org/Acrobat/If-I-am-assassinated-by-Shaheed-Bhutto.pdf

Bodansky, B. (1998), "Pakistan's Islamic Bomb", Houston: Freeman Center for Strategic Studies, [Online: web] Accessed on 19 April 2016, URL: http://www.freeman.org/m_online/jul98/bodansky.htm.

Bolan, C. J. (2013), "Dealing With Iran: The Iranian Nuclear Debate: More Myths Than Facts", *Parameters,* 43(2): 77-88

Bonney, R. (2012), "General Editor's Introduction", in Rizwana Abbasi, *Pakistan and the New Nuclear Taboo: Regional Deterrence and the International Arms Control Regime*, Bern: Peter Lang.

Bowen, W. and Moran, M. (2014), "Iran's Nuclear Programme: A Case Study in Hedging?", *Contemporary Security Policy*, 35 (1): 26-52.

Buchta, W. (2000), *Who rules Iran?: the structure of power in the Islamic Republic*, Washington: The Washington Institute for Near East Policy and the Konrad Adenauer Stiftung.

Byman, D. L. and Pollack, K. M. (2001), "Let Us Now Praise Great Men: Bringing the Statesman Back In", *International Security*, 25 (4):107-146.

Carr, E. H. (1946), *The Twenty Years' Crisis 1919-1939: An Introduction To The Study of International Relations*, London: Macmillan & Co. Ltd.

Cavanaugh, W. T (1999), *The Myth of Religious* Violence, New York: Oxford University Press

Cavanaugh, W. T (2013), "What is Religion?", in *Religion and International Relations: A Primer for Research*, The Report of the Working Group on International Relations and Religion of the Mellon Initiative on Religion Across the Disciplines, University of Notre Dame, P. 56-67

Chafetz, G. (1993) "The End of the Cold War and the Future of Nuclear Nonproliferation: An Alternative to the Neo-realist Perspective," *Security Studies*, 2: 127-158.

Chakma, B. (2002), "Road to Chagai: Pakistan's Nuclear Programme, Its Sources and Motivations", *Modern Asian Studies*, 36 (4): 871-912

Cirincione, J. et al. (2005), "Pakistan" in *Deadly Arsenals: Nuclear, Biological, and Chemical Threats*, Washington: Carnegie Endowment for International Peace.

Clarke, M. (2013), "Iran as a 'pariah' nuclear aspirant", *Australian Journal of International Affairs*, 67 (4): 491-510.

Cohen, S. P. (2004), *The Idea of Pakistan*, Washington, D.C.: Brookings Institution Press.

Collier, R. (2003), "Nuclear weapons unholy, Iran says / Islam forbids use, clerics proclaim", *S. F. Gate*, 31 October 2003, [Online: web] Accessed on 25 March 2016, URL: http://www.sfgate.com/news/article/Nuclear-weapons-unholy-Iran-says-Islam forbids-2580018.php.

Dajani, A. (2011), "Islamic Nationalism VS Islamic Ummatism/ *al-Ummatya*: Conceptualizing Political Islam (Ummatism/ *al-Ummatya*) (Ummatist/

Ummatawee) (Ummatists / *Ummatawyon*)", Kings College London, [Online: web] Accessed on 15 April 2016, URL: https://www.soas.ac.uk/lmei/events/ ssemme/file67881.pdf.

Dark, K. R. (2000), *Religion and International Relations*, Britain: Macmillan Press Ltd.

Desch, M. C. (2013), "The Coming Reformation of Religion in International Affairs? The Demise of the Secularization Thesis and the Rise of New Thinking about Religion" in *Religion and International Relations: A Primer for Research*, The Report of the Working Group on International Relations and Religion of the Mellon Initiative on Religion Across the Disciplines, University of Notre Dame, p. 14-54

Deutsch, J. M. (1992), "The New Nuclear Threat", *Foreign Affairs*, 71(41): 124-125.

Eisenstadt, M. (2011), "Religious Ideologies, Political Doctrines, and Nuclear Decisionmaking", in Michael Eisenstadt and Mehdi Khalaji (2011), *Nuclear Fatwa Religion and Politics in Iran's Proliferation Strategy*, Washington: Washington Institute for Near East Policy

Epstein, W (1977), "Why States Go -- And Don't Go – Nuclear", *Annals of the American Academy of Political and Social Science*, Vol. 430: 16-28

Erickson, J. L and Way, C. (2011), "Membership has its Privileges: Conventional Arms and Influence within the Nuclear Non-Proliferation Treaty" in Robert Rauchhaus et al. (eds.) *Causes and Consequences of Nuclear Proliferation*, London, New York: Routledge.

Farmer, M. L. (2005), *Why Iran Proliferates*, Thesis, Monterey, California: Naval Postgraduate School.

Finnemore, M. (1996), "Constructing Norms of Humanitarian Intervention" in in Peter J. Katzenstein (ed.) *The Culture of National Security: Norms and Identity in World Politics*, New York: Columbia University Press

Fitzpatrick, M. (2015), "Saudi Arabia, Pakistan and the Nuclear Rumour Mill", *Survival*, 57(4):105-108.

Fox, J and Sandler, S. (2004), *Bringing Religion into International Relations*, New York: Palgrave Macmillan

Frey, K. (2006), "Nuclear Weapons as Symbols: The Role of Norms in Nuclear Policy Making", *IBEI Working Papers*, IBEI: De EstaEdición.

Gorski, P. S. and Dervişoğlu, G. T. (2013), "Religion, Nationalism, and International Security: Creation Myths and Social Mechanisms", in Chris Seiple, Dennis R. Hoover and Pauletta Otis(eds.), *The Routledge Handbook of Religion and Security*, London and New York: Routledge

Gut, J. L. (2013), "Religion and Public Opinion on Security: A Comparative Perspective", in Chris Seiple, Dennis R. Hoover and Pauletta Otis(eds.), *The Routledge Handbook of Religion and Security*, London and New York: Routledge

Habibzadeh, T. (2014), "Nuclear Fatwa and International Law", Iranian Review of Foreign Affairs, 5(3): 149-177.

Harris, S. and Aid, M.M. (2013), "Exclusive: CIA Files Prove America Helped Saddam as He Gassed Iran: The U.S. knew Hussein was launching some of the worst chemical attacks in history -- and still gave him a hand", *Foreign Policy* 26 August 2013, [Online: web] Accessed on 23July 2017, URL: http://foreignpolicy.com/2013/08/26/exclusive-cia-files-prove-america-helped-saddam-as-he-gassed-iran/.

Hashmi, S. H (2004), "Islamic Ethics and Weapons of Mass Destruction: An Argument for Nonproliferation", in Sohail H. Hashmi and Steven P. Lee, (eds.) (2004), *Ethics and Weapons of Mass Destruction: Religious and Secular Perspectives*, Cambridge, etc.: Cambridge University Press.

Hashmi, S. H and Lee, S. P (2004), *Ethics and Weapons of Mass Destruction: Religious and Secular Perspectives*, Cambridge, New York, Melbourne, Madrid, Cape Town, Singapore, Sao Paulo: Cambridge University Press

Herrington, L. M. et al. (2015), *Nations under God The Geopolitics of Faith in the Twenty-First Century*, Bristol, England: E-International Relations

Hilali, A. Z. (2002), "Pakistan's Nuclear Deterrence: Political And Strategic Dimensions", Centre for Strategic Research, [Online: web] Accessed on 24 April 2016, URL: http://sam.gov.tr/wp-content/uploads/2012/01/A.Z.-Hilali.pdf

Hitchcock, M (2007), *The Apocalypse of AhmadINejad: The Revolution of Iran's Nuclear prophet*, Colorado: Multnomah Books

Hoodbhoy, P. (1993), "Myth-building: The 'Islamic' Bomb". *The Bulletin of the Atomic Scientists*, 49 (5): 42-49.

Hoodbhoy, P. (1995), "Nuclear Myths and Realities", in Zia Mian (ed.), *Pakistan's Atomic Bomb and the Search for Security*, Lahore: Gautam Publishers.

Hoodbhoy, P. (2010), "Can the Islamic Bomb Become a Reality", *Palestine-Israel Journal*, 16 (34): 77-81.

Hoodbhoy, P. (2013), "Iran, Saudi Arabia, Pakistan and the 'Islamic Bomb'", Chapter 7 in Pervez Hoodbhoy (ed.), *Confronting the Bomb: Pakistani and Indian Scientists Speak Out*, Oxford: Oxford University Press.

Hoodbhoy, P. and Mian, Z. (2014), "Nuclear fears, hopes and realities in Pakistan", *International Affairs*, 90 (5): 1125–1142.

Hymans, J.E.C. (2006), *The Psychology of Nuclear Proliferation: Identity, Emotions, and Foreign Policy*, Cambridge: Cambridge University Press.

Hymans, J.E.C. (2010), "Theories of Nuclear Proliferation", *The Nonproliferation Review*, 13 (3): 455-465.

ISIS Report (2013), "Iran's Nuclear History from the 1950s to 2005", Institute for Science and International Security, April 17, 2013, [Online: web] Accessed on 25 March 2016. URL: http://www.isisnucleariran.org/assets/pdf/Iran_Nuclear_History.pdf.

Islam Today (2010), "Iran' Supreme Leader: Nuclear Weapons are Against the Teachings of Islam", 20 February 2010, [Online: web] Accessed on 23 Jan. 2015 URL: http://en.islamtoday.net/print/3519.

Jafarzadeh, A. (2007), *The Iran Threat, President Ahmadinejad And The Coming Nuclear Crisis*, New York: Palgrave Macmillan

Jepperson, R. L. et al (1996), "Norms, Identity, and Culture in National Security" in Peter J. Katzenstein (ed.) *The Culture of National Security: Norms and Identity in World Politics*, New York: Columbia University Press

Johnson, J. L. et al. (2009), *Strategic Culture and Weapons Of Mass Destruction*, New York: Palgrave Macmillan

Johnson, J. T (2011), *Ethics and the Use of Force: Just War in Historical Perspective*, Farnham: Ashgate Publishing Limited

Johnston, D. and Samson, C. (1994), *Religion: The Missing Dimension of Statecraft* Oxford: Oxford University Press

Jones, P. (2012), "Learning to Live with a Nuclear Iran", *The Nonproliferation Review*, 19 (2): 197-217.

Kartchner, K. M. (2009), "Strategic Culture and WMD Decision Making", in Jeannie L. Johnson, Kerry M. Kartchner, and Jeffrey A. Larsen (eds.) *Strategic Culture and Weapons of Mass Destruction*, New York: Palgrave Macmillan

Katzenstein, P. J. (1996), *The Culture of National Security: Norms and Identity in World Politics*, New York: Columbia University Press

Kaushik, B. M and Mehrotra, O. N. (1980), *Pakistan's Nuclear Bomb*, New Delhi: Sopan Publishing House.

Kelsay, J. (2006), "Islamic Tradition and the Justice of War," in T. Brekke, ed., *The Ethics of War in Asian Civilizations*, New York: Routledge: 81-110.

Kerr, P. K. And Nikitin, M. B. (2016), "Pakistan's Nuclear Weapons", Congressional Research Service Report, [Online: web] Accessed on 19 April 2016, URL: https://www.fas.org/sgp/crs/nuke/RL34248.pdf

Khalaji, M. (2011), "Shiite Jurisprudence, Political Expediency, and Nuclear Weapons', in Michael Eisenstadt and Mehdi Khalaji (2011), *Nuclear Fatwa Religion and Politics in Iran's Proliferation Strategy*, Washington: Washington Institute for Near East Policy.

*Khamenei, A. A. (2012), "Nuclear weapon is Haraam; Nuclear energy is a right", 30 August 2012, [Online: web] Accessed on 8 April 2016, URL: http://english. khamenei.ir/news/2270/Nuclear-weapon-is-Haraam-Nuclear-energy-is-a-right.

*Khamenei, A.A. (2012a), Leader's Inaugural Speech at the 16th Non-Aligned Summit + Video", 30 August 2012, [Online: web] Accessed on 23July 2017, URL: http://english.khamenei.ir/news/1668/Leader-s-Inaugural-Speech-at-the-16th-Non-Aligned-Summit-Video

Khan, F. H. (2005), "Comparative Strategic Culture: The Case of Pakistan", *Strategic Insights*, 4 (10).

Khan, F. H. (2012), *Eating Grass: The Making of the Pakistani bomb,* Stanford: Stanford University Press, Reprinted in 2013, New Delhi: Cambridge University Press India Pvt. Ltd.

Khan, S. (2010), *Iran and Nuclear Weapons Protracted conflict and proliferation*, Oxon and New York: Routledge.

Khan, W. S. (2010), "There's a Reason Why They Call It 'The Islamic Bomb'", [Online: web] Accessed on 21 April 2016, URL: https://wajskhan.wordpress. com/2010/04/16/desi-nukespeak-political-discourse-behind-the-divine-myth of pakistan%E2%80%99s-islamic-bomb/.

Khan, Z. (2015), *Pakistan's Nuclear Policy: A Minimum Credible Deterrence,* London and New York: Routledge

Kibaroglu, M. (2006), "Good for the Shah, Banned for the Mullahs: The West and Iran's Quest for Nuclear Power", *Middle East Journal*, 60 (2): 207-232

Krasner, S.D. (1983), "Structural Causes and Regime Consequences: Regimes as Intervening Variables," in Stephen D. Krasner (ed.) *International Regimes*, Ithaca, New York: Cornell University Press.

Larssen, R. M (2011), *Islam And The Bomb: Religious Justification For And Against Nuclear Weapons*, Belfer Centre for Science and International Affairs: Harvard Kennedy School.

Lavoy, P. R. (1993), "Nuclear Myths and the Causes of Nuclear Proliferation", *Security Studies*, 2 (3/4): 192-212

Lavoy, P. R. (2006), "Nuclear Proliferation over the Next Decade Causes, Warning Signs, and Policy Responses", *Nonproliferation Review*, 13 (3): 433-454.

Lavoy, P. R. (2006), "Pakistan's Strategic Culture", Defense Threat Reduction Agency, [Online: web] Accessed on 17 April 2016, URL:https://fas.org/irp/agency/dod/dtra/pakistan.pdf.

Linzer, D. (2005), "Past Arguments Don't Square With Current Iran Policy", *Washington Post*, 27 March 2005, [Online: web] Accessed on 23July 2017, URL: http://www.washingtonpost.com/wp-dyn/articles/A3983-2005Mar26.html.

Long, W.J and Grillot, S.R. (2000), "Ideas, beliefs, and nuclear policies: The cases of South Africa and Ukraine", *The Nonproliferation Review*, 7 (1): 24-40.

Marsden, L. and Savigny, H (2013), "Religion, Media, and Security" in Chris Seiple, Dennis R. Hoover and Pauletta Otis(eds.), *The Routledge Handbook of Religion and Security*, London and New York: Routledge

Maurer, C. L. (2014), *Iran: The Next Nuclear Threshold State?*, Thesis, Monterey, California: Naval Postgraduate School

Mayer, C. C. (2004), *National Security to Nationalist Myth: Why Iran Wants Nuclear Weapons*, Thesis, Monterey, California: Naval Postgraduate School

Mearsheimer, J. J. (2001), *The Tragedy of Great Power Politics*, New York and London: W. W. Norton & Company.

Meyer, S. M. (1984), *The Dynamics of Nuclear Proliferation*, Chicago: University of Chicago Press.

Morgenthau, H. J. (1948), *Politics among Nations: The Struggle for power and Peace*, New York: Alfred A. Knopf.

Mousavian, H. (2012), "The Iranian Nuclear Dispute: Origins and Current Options", [Online: web] Accessed on 23July 2017, URL: http://www.princeton.edu/sgs/faculty-staff/seyed-hossein-mousavian/Iranian-Nuclear-Dispute-Origins.pdf.

Mousavian, H. and Afrasiabi, K. (2012), "Eight Reasons Why Waltz Theory on Nuclear Iran is Wrong", *Al Monitor*.

Mousavian, S. H. (2012), "20 Reasons Iran is not after Nuclear Bomb", *IPPNW - International Physicians for the Prevention of Nuclear War*, Berlin.

Mousavian S.H. (2012a), The Iran Nuclear Dilemma: The Peaceful Use of Nuclear Energy and NPT's Main Objectives". *EU Non-Proliferation Consortium*, [Online: web] Accessed on 23July 2017, URL: http://www.nonproliferation. eu/web/documents/backgroundpapers/mousavian.pdf.

Mousavian S. H. (2012b), "Iran, the US and Weapons of Mass Destruction", *Survival*, 54 (5): 183-202.

Mousavian, S. H. (2013), "Globalizing Iran's fatwa against Nuclear weapons", *Survival: Global Politics and strategy*, 55 (2): 147-162.

Mousavian, S. H (2017), "Understanding Iranian threat perceptions", *Al-Monitor*, 14 July 2017, [Online: web] Accessed on 23July 2017, URL: http://www. al-monitor.com/pulse/originals/2017/07/iran-threat-perceptions-regime-change-regional-dialogue.html

Neill, B. O. (2002), "Nuclear Weapons and the Pursuit of Prestige", Available, [Online: Web], accessed 10, October, 2015, URL: http://papers.ssrn.com/sol3/papers.cfm?abstract_id=887333

Nye, J. S. (1986), *Nuclear Ethics*, London: Collier Macmillan Publishers.

Nye, J. S. (2004), *Soft Power: The Means to Success in World Politics*, New York: Public Affairs.

Palit, D. K. and Namboodiri, P. K. S. (1979), *Pakistan's Islamic Bomb*, New Delhi: Vikas Publishing House Pvt. Ltd.

Patrikarakos, D. (2012), *Nuclear Iran: The Birth of an Atomic State*, New York: I. B. Tauris.

Patterson, E. (2013), "Religion, War, and Peace: Leavening the Levels of Analysis", in Chris Seiple, Dennis R. Hoover and Pauletta Otis(eds.), *The Routledge Handbook of Religion and Security*, London and New York: Routledge

Paul, T. V. (2000), *Power versus Prudence: Why Nations Forgo Nuclear Weapons*, London, Ithaca: McGill-Queen's University Press.

Paul, T. V. (2009), *The Tradition of Non-Use of Nuclear Weapons*, Stanford Security Studies, Stanford and California: Stanford University Press.

Paul, T. V. et al. (eds.) (2014), *Status in World Politics*, Cambridge and New York: Cambridge University Press.

Porter, G. (2014), "When the Ayatollah Said No to Nukes", *Foreign Policy*, 16 October 2014, [Online: web] Accessed on 20 march 2016 URL: http://foreignpolicy.com/2014/10/16/when-the-ayatollah-said-no-to-nukes/.

Qadiri, M. T. (2010), *Fatwa on Terrorism and Suicide Bombing*, Minhaj-ul-Quran International

Qazi, S. H. (2013), "Making the Bomb: Pakistan's Nuclear Journey", Review Essay on Feroz H. Khan, *Eating Grass: The Making of the Pakistani Bomb*, *World Affairs*, 176 (2): 88-92.

Quinlan, M. (2009), *Thinking About Nuclear Weapons Principles, Problems, Prospects*, New York: Oxford University Press.

Rauchhaus, R. et al. (eds.) *Causes and Consequences of Nuclear Proliferation*, London, New York: Routledge.

Read, R. (2017), "Mattis: Iran Needs Regime Change for Relations to Improve with US", *Daily Caller*, 10 July 2017, [Online: web] Accessed on 23July 2017, URL: http://dailycaller.com/2017/07/10/mattis-iran-needs-regime-change-for-relations-to-improve-relations-with-us/.

Reuters (2012), "FACTBOX-Tehran Research Reactor", 15 February 2012, [Online: web] Accessed on 23July 2017, URL: http://www.reuters.com/article/iran-nuclear-reactor-idAFL5E8DF2I720120215.

Rizvi, H. A. (2002), "Pakistan's Strategic Culture", Chapter 12 in Michael R. Chambers (Ed.) *South Asia in 2020: Future Strategic Balances and Alliances* Carlisle Barracks, PA: Strategic Studies Institute, U.S. Army War College.

Rowberry, A. (2013), "Sixty Years of "Atoms for Peace" and Iran's Nuclear Program", [Online: web] Accessed on 23July 2017, URL: https://www.brookings.edu/blog/up-front/2013/12/18/sixty-years-of-atoms-for-peace-and-irans-nuclear-program/.

Sagan, S. D. (1995), "More Will be Worse" in Scott D. Sagan and Kenneth N. Waltz (eds.) *The Spread of Nuclear Weapons: A Debate Renewed*, New York and London: W.W. Norton& Company.

Sagan, S. D. (1996), "Why Do States Build Nuclear Weapons?: Three Models in Search of a Bomb", *International Security*, 21 (3): 54-86.

Sagan, S. D. (2004), "Realist perspectives on Ethical Norms and Weapons of Mass Destruction", in Sohail H. Hashmi and Steven P. Lee, (eds.) (2004), *Ethics and Weapons of Mass Destruction: Religious and Secular Perspectives*, Cambridge, etc.: Cambridge University Press.

Sahimi, M. (2010), "Iran's Uranium Enrichment Program (Part I)", [Online: web] Accessed on 23July 2017, URL: http://www.pbs.org/wgbh/pages/frontline/tehranbureau/2010/03/irans-uranium-enrichment-program-part-i.html.

Sandal, N. A. and James, P. (2010), "Religion and International Relations Theory: Towards a mutual understanding", *European Journal of International Relations*, 17(1): 3–25

Sasikumar, K. (2006), *Regimes at Work: The Nonproliferation Order and Indian Nuclear Policy*, Ph.D Thesis, Cornell University.

Schmidt, O. (2008), *Understanding &Analyzing Iran's Nuclear Intentions: Testing Scott Sagan's Argument of "Why do States build Nuclear Weapons*, Lancaster University

Sciolino, E. (1998), "The World: Buzz Words; Who's Afraid of the Islamic Bomb?", *The New York Times*, 7 June 1998.

Shah, T. S. (2013), "Religion and International Relations: Normative Issues", in *Religion and International Relations: A Primer for Research*, The Report of the Working Group on International Relations and Religion of the Mellon Initiative on Religion Across the Disciplines, University of Notre Dame, P. 87-102

SNIE 4-1-74 (1974), "Prospects for Furthur Proliferation of Nuclear Weapons", Special National Intelligence Estimate, [Online: web] Accessed on 23July 2017, URL: http://nsarchive.gwu.edu/NSAEBB/NSAEBB240/snie.pdf.

Sokolski, H. and Clawson, P. (2005), *Getting Ready for A Nuclear-Ready Iran*, USA: The Strategic Studies Institute

Sollingen, E. (1994), "The Political Economy of Nuclear Restraint", *International Security*, Vol. 19 (2): 126-169

Tagma, H. M. and Uzun, E. (2012), "Bureaucrats, Ayatollahs, and Persian Politics: Explaining the Shift in Iranian Nuclear Policy", *The Korean Journal of Defense Analysis*, 24 (2): 239–264.

Tannenwald, N. (2007), *The Nuclear Taboo: The United States and the Non-Use of Nuclear Weapons since 1945*, Cambridge and New York: Cambridge University Press

Thayer, B.A. (1995), "The Causes of Nuclear Proliferation and the Non-proliferation Regime", *Security Studies*, Vol. 4(3): 463-519

Thomas, S. (2000), "Religion and International Conflict", in K. R. Dark (ed.), *Religion and International Relations*, Britain: Macmillan Press Ltd.

VOA (*Voice of America English News*), (2010), "Iran's Supreme Leader Khamenei Says Islam Opposes Nuclear Weapons", February 18, 2010, [Online: web] Accessed on 23 Jan. 2016 URL: http://www.voanews.com/articleprintview/112715.html.

Waltz, K. (1954), *Man, the state, and war: a theoretical analysis*, New York: Columbia University Press.

Waltz, K. (1981), "The Spread of Nuclear Weapons: More May Better," *Adelphi Papers*, Number 171, London: International Institute for Strategic Studies.

Waltz, K. N. (2012), "Why Iran Should Get the Bomb: Nuclear Balancing Would Mean Stability", *Foreign Affairs*, July-August 2012, [Online: web] Accessed on 24 November 2015 URL: https://www.foreignaffairs.com/articles/iran/2012-06-15/why-iran-should-get-bomb.

Weissman, S. and Krosney, H. (1981), *The Islamic Bomb*, New York: Times Books, Reprinted in India in 1983, New Delhi: Orient Paperbacks.

Wendt, A. (1992), "Anarchy is what States Make of it: The Social Construction of Power Politics", *International Organization*, 46 (2): 391-425

White, T. O. (1996), "Is There A Theory Of Nuclear Proliferation? An Analysis of the Contemporary Debate", *The Nonproliferation Review*, 13 (3): 43-60.

Yasmeen, S. (2001), "Is Pakistan's Nuclear Bomb An Islamic Bomb?", *Asian Studies Review*, 25 (2): 201-215.

Zakaria, F. (2009), "What You Know About Iran is Wrong", *Newsweek*, 22 May 2009, [Online: web] Accessed on 24 March 2016 URL: http://www.newsweek.com/zakaria-what-you-know-about-iran-wrong-80049.

Zarif, M. J. (2016), "Why Iran is building up its defenses", *The Washington Post*, 20 April 2016, [Online: web] Accessed on 30 July 2017, URL: https://www.washingtonpost.com/opinions/zarif-what-critics-get-wrong-about-iran-and-the-nuclear-agreement/2016/04/20/7b542dee-0658-11e6-a12f-ea5aed7958dc_story.html?utm_term=.7d6f4f3c33d3.

Index

About Authors

Dr. Shameer Modongal teaches West Asian Studies at the University of Kerala. He completed PhD from the Centre for International Politics, Organization, Diplomacy and Disarmament (CIPOD) at Jawaharlal Nehru University (JNU), New Delhi. His book, Islamic Perspectives on International Conflict Resolution, was published by Routledge in 2022. In 2021, Vij Books published his edited book Counterterrorism and Global Security: Genesis, Responses and Challenges. Additionally, his works have appeared in many international peer-reviewed journals like Bandung: Journal of the Global South, Cogent Social Sciences, Defense & Security Analysis, Defence Studies, Digest of Middle East Studies, Insight Turkey, International Studies Review, Human Rights Review, Journal of Asian Security and International Affairs, Politics, Religion & Ideology, Political Studies Review and Strategic Studies. He is also serving as an editor/reviewer of many international journals.

Ambassador Seyed Hossein Mousavian is a Middle East Security and Nuclear Policy Specialist at Princeton University and a former spokesperson for Iran's nuclear negotiators. His book on nuclear crisis, The Iranian Nuclear Crisis: A Memoir, was published in 2012 by Carnegie Endowment for International Peace. A second book, Iran and the United States: An Insider's View on the Failed Past and the Road to Peace, was released in May 2014. Routledge published his new book, A Middle East Free of Weapons of Mass Destruction, in May 2020.

www.ingramcontent.com/pod-product-compliance
Lightning Source LLC
Chambersburg PA
CBHW021715210326
41599CB00013B/1658